SpringerBriefs in Applied Sciences and Technology

More information about this series at http://www.springer.com/series/8884

T.M.V. Suryanarayana · P.B. Mistry

Principal Component Regression for Crop Yield Estimation

 Springer

T.M.V. Suryanarayana
Water Resources Engineering
  and Management Institute
The Maharaja Sayajirao University
  of Baroda
Vadodara, Gujarat
India

P.B. Mistry
Vadodara, Gujarat
India

ISSN 2191-530X   ISSN 2191-5318   (electronic)
SpringerBriefs in Applied Sciences and Technology
ISBN 978-981-10-0662-3   ISBN 978-981-10-0663-0   (eBook)
DOI 10.1007/978-981-10-0663-0

Library of Congress Control Number: 2016934955

This Springer imprint is published by Springer Nature
The registered company is Springer Science+Business Media Singapore Pte Ltd.

Preface

Since the start of the present century, climate change has been the topic of discussion in varied fields. In the same line, the topic gained its importance in the areas of water resources and agriculture. The climate with its various important variables, i.e., climatological variables, has a direct or indirect impact on agricultural production. There is a need to study the effect of climatological variables and their dominance in crop yield estimation. Downscaling techniques in general and statistical downscaling method in particular and principal component analysis in detail are discussed. This book will be helpful to the students and researchers, who are starting their works on climate and agriculture with a special focus on estimation models. The flow of chapters takes the readers in a smooth path, starts in understanding climate and weather and impact of climate change, and gradually proceeds towards downscaling techniques and then finally towards development of principal component regression models, and applies the same for the crop yield estimation.

Chapter 1 starts with differentiation of climate and weather and then discussions on climate change and its impact on global context, and mainly on agriculture, are highlighted. The various climatological parameters affecting crop yield are then discussed. This is followed by a brief description of downscaling techniques and their application. This chapter concludes with the brief introduction to multiple linear regression and principal component analysis, followed by the objectives of study.

Chapter 2 explores the transfer function in detail, with multiple linear regressions, and principal component analysis. Furthermore, it contains a brief description of various types of regression and emphasizes on the principal component analysis and the calculations of principal components (PCs) in detail.

Chapter 3 covers the various research works carried out in the field of analyzing climatic variability induced due to climate change scenarios and its impact on agriculture. The reviews of works are classified into four domains, viz., climate change, downscaling techniques, multiple linear regression, and principal component analysis and principal component regression.

Chapter 4 contains a brief overview of agroclimatic zones of India followed by the sub agroclimatic zones of Gujarat. Thereafter, a brief description of the study area and the data required for the study are provided.

Chapter 5 gives an overview on the methodology to predict the yield of cotton using multiple linear regression and principal component regression, which is followed by the description of the performance indices to estimate the best model. The methodology of MLR and PCR models to predict the crop yield for the study area, considering the climatological parameters as input and yield as output, is discussed.

Chapter 6 illustrates the results and analysis of the study. The MLR models are developed and the performance indices for the MLR model are analyzed. This is followed by the development and evaluation of PCR model using the performance indices. The outcomes are statistically analyzed and their accuracy is assessed and discussed during training and validation.

Chapter 7 summarizes the results obtained during the study, the comparisons of the results by MLR and PCR are focused, and the conclusions of the study are drawn.

Acknowledgment

It was indeed a nice experience in working for publishing with Springerbriefs. The authors are very much thankful to Dr. Loyola D'Silva, Publishing Editor, Springer, and Ms. Amudha Vijayarangan, Project Coordinator (Books), Springer for their continuous support during the process of publication. We are also thankful to Tom Steendam and all other editorial members and staff of Springer, with whose support, this has been published.

We are also thankful to the reviewers for sparing their valuable time in reviewing the content of this book.

Last but not least, we are also thankful to our family members, whose continuous support and encouragement motivated the zeal in writing this book.

T.M.V. Suryanarayana
P.B. Mistry

Contents

About the Authors

Dr. T.M.V. Suryanarayana is Associate Professor and a recognized Ph.D. Guide in Water Resources Engineering and Management Institute, The M.S. University of Baroda. He is Executive Committee Member of Indian Water Resources Society, Secretary and Treasurer of Gujarat Chapter of Association of Hydrologists of India and Joint Secretary of Indian Society of Geomatics_Vadodara Chapter. He has published 75 research papers in various international/national journals/seminars/conferences/symposiums.

Mr. P.B. Mistry has obtained B.E. (Civil-Irrigation Water Management) and M.E. (Civil) in Water Resources Engineering from The M.S. University of Baroda and is currently a freelancer from Vadodara working in the areas of water resources and irrigation water management. He is a life member of Indian Society of Geomatics and Indian Society for Hydraulics.

List of Figures

List of Tables

Chapter 1
Introduction

Abstract This chapter starts from differentiating climate and weather and then climate change and its impact on global context and mainly on agriculture. The various climatological parameters affecting crop yield are then discussed. This is followed by the slight description of downscaling techniques and its application. This chapter concludes with the brief introduction of multiple linear regression (MLR) and principal component analysis, followed by the objectives of the study.

Keyword Climate and Weather · Climatological parameters · Crop yield · Downscaling · MLR · PCR

1.1 Climate and Weather

Climate is generally defined as the average weather, and so, climate change and weather are intertwined. Observations can show that there have been changes in weather, and it is the statistics of changes in weather over time that identifies climate change. While weather and climate are closely related, there are important differences. A common confusion between weather and climate arises, when scientists are asked how they can predict climate 50 years from now when they cannot predict the weather a few weeks from now. The chaotic nature of weather makes it unpredictable beyond a few days. Projecting changes in climate (i.e., long-term average weather) due to changes in atmospheric composition or other factors is a very different and much more manageable issue. There are always hot and cold extremes, although their frequency and intensity change as climate changes. But when weather is averaged over space and time, the fact that the globe is warming emerges clearly from the data (Climate Change 2007: Working Group I: The Physical Science Basis, FAQ 1.2).

Climate is one of the key parameters in the earth's environment. Climate is usually defined as the average weather and in broad sense, it is the statistical description in terms of the mean and variability of relevant quantities over a period of time, ranging from months to thousands or millions of years. Human activities

T.M.V. Suryanarayana and P.B. Mistry, *Principal Component Regression for Crop Yield Estimation*, SpringerBriefs in Applied Sciences and Technology, DOI 10.1007/978-981-10-0663-0_1

that could possibly result in change of the climate include the emission of gases into the atmosphere, industrial activities, development of extensive cities, pollution of water ways and cities, creation of thousands of dams and lakes, conversion of grassland or forest to cropland, agricultural activities.

The average global temperature rises by 0.74 °C over the last hundred years (1906–2005), with more than half of these rises, 0.44 °C, in the last 25 years. Most of the warming over the last 50 years is very likely to have been caused by anthropogenic increases in Green House Gases (GHGs). Since 1750, atmospheric concentrations of GHGs have increased significantly. Carbon dioxide has increased by 31 %, Methane by 151 %, and Nitrous oxide by 17 %. Higher carbon dioxide concentration is caused due to burning of fossil fuels (coal, oil, and natural gas) and deforestation (Gautam 2010).

1.2 Climate Change

Climate change is a long-term shift or alteration in the climate of a specific location, region, or the entire planet. The shift is measured by changes in some or all of the features associated with average weather, such as temperature, wind patterns, and precipitation. It can involve both changes in average weather conditions and changes in how much the weather varies about these averages. "Climate Change" is distinguished from "Climate Variability" by the persistence of the change over time, so that a measurable difference is observed between two periods of time.

At the global scale, climate change occurs in response to a change in the amount of energy flowing into or out of the earth's climate system. This occurs when something alters either the amount of the sun's radiation absorbed by the earth's atmosphere and surface, or the amount of heat radiation emitted from the earth's surface and atmosphere to space. The climate system responds to this imbalance in energy input versus output by warming or cooling, until a radiation energy balance is restored. Since the factors that cause the initial change in the energy balance push or "force" the climate to change, these factors are generally referred to as "climate forcing." Colloquially, positive forcing are often referred to as "warming factors" while negative forcing are called "cooling factors." Climate forcing can be a natural phenomena or canaries from human activities. The factors that affect regional climate change are much more complex. That is because, in addition to being affected by global climate change, regional climates are also affected by a myriad of other factors operating on smaller time and space scales, and by changes in wind and ocean patterns due to internal fluctuations of the climate system (A.2: FAQ on Science of Climate Change).

Climate change is a complex problem that has increased the need for an integrated, multisectoral, and multidisciplinary response. Apart from the normal water domain, decision-makers in other spheres, i.e., finance, trade, energy, housing, regional planning, agriculture must use and consume water efficiently. Sustainable management and development of water resources will play a pivotal role in preparing societies' ability to adapt to climate change in order to increase resilience and achieve development goals. This calls for policy and governance shifts, investments and changes in the way water concerns are addressed in development strategies and budgets (Climate Change Adaptation: The Pivotal Role of Water).

1.3   Impact of Climate Change in Global Context

Impact of climate change can be categorized through positive and negative aspects. Less chilly winters and greenery in high altitudinal areas can be considered as some positive impacts due to global temperature rise. However, the adverse (negative) impacts are seen very high in compared to the positive impacts. Some of the adverse impacts caused by climate change especially related to water in global and Asian context can be categorized as follows (Gautam 2010):

- It is seen that the global temperature has been on the rising trend since mid-twentieth century.
- Hot days and hot nights have become more frequent in most parts of the world.
- Due to rising temperature, it causes abrupt glacier ablation. The formation of lakes is occurring as glaciers retreat from several steep mountain ranges, including the Himalayas. These lakes thus have a high potential for glacial lake outburst floods (GLOFs).
- Climate change has induced changes in surface and groundwater systems. At the global scale, there is evidence of a broadly coherent pattern of change in annual runoff.
- In some regions like China, higher latitude regions experience an increase in runoff while West Africa, southern Europe, and southern Latin America experience the decrease in runoff (Bates et al. 2008).
- Many natural systems on all continents and oceans are affected due to global warming.
- Diseases related to warming and mosquito problems are seen even in the high altitudinal regions.
- Changes in water quantity and quality due to climate change, effects on food production, leading to decrease in food security.
- It is found that the rate of sea-level rise during the twentieth century was about 10 times higher than average rate during the last 3000 years.
- Sea-level rise is projected to extend areas of salinization of groundwater, resulting in a decrease of freshwater availability for humans and ecosystems in coastal areas (Bates et al. 2008).

The observed effects of climate change and its observed/possible impacts on water services in global perspectives are shown in Table 1.1.

Table 1.1 Observed effects of climate change and its observed/possible impacts on water services in global perspectives

Observed effect	Observed/possible impacts
Increase in atmospheric temperature	Reduction in water availability in basins fed by glaciers that are shrinking, as observed in some cities along the Andes in South America
Increase in surface water temperature	Reductions in dissolved oxygen content, mixing patterns, and self-purification capacity Increase in algal blooms
Sea-level rise	Salinization of coastal aquifers
Shifts in precipitation patterns	Changes in water availability due to changes in precipitation and other related phenomena (e.g., groundwater recharge, evapotranspiration)
Increase in interannual precipitation variability	Increase the difficulty of flood control and reservoir utilization during the flooding season
Increased evapotranspiration	Water availability reduction Salinization of water resources Lower groundwater levels

1.4 Impact of Climate Change on Agriculture

Climate is an important factor of agricultural productivity. Concerns have been expressed by many organizations and others regarding the potential effects of climate change on the same. Interest in this matter has motivated a substantial body of research on climate change and agriculture over the past decade. Climate change is expected to affect agricultural and livestock production, hydrologic balances, input supplies, and other components of agricultural systems.

Climate change is caused by the release of "greenhouse" gases into the atmosphere. These gases accumulate in the atmosphere and result in global warming. The changes in global climate related parameters such as temperature, precipitation, soil moisture, and sea level are observed. However, the reliability of the predictions on climate change is uncertain. There are no hard facts about what will definitely be the result of increase in the concentration of greenhouse gases within the atmosphere and no firm timescales are known. Agriculture is one sector, which is important to consider in terms of climate change. The agriculture sector will contribute to climate change, and also be affected by the changing climate.

The climate change effects on agriculture will differ across the world. To determine how the climate change affects agriculture is complex, and varieties of effects are likely to occur. Changes in temperature as well as changes in rainfall patterns and the increase in carbon dioxide levels projected to accompany climate change will have important effects on global agriculture, especially in the tropical regions. It is expected that crop productivity will alter due to the changes in climate/weather events and changes in patterns of pests and diseases. The suitable land areas for cultivation of key staple crops could undergo geographic shifts in response to climate change (Aydinalp and Cresser 2008).

1.5 Climatological Parameters Affecting Crop Yeild

The climatological parameters affecting the crop yield are mainly considered as maximum and minimum temperature, relative humidity, wind speed, and sunshine hours. The importance of the same is discussed hereafter.

1.5.1 Maximum and Minimum Temperature

The degree of hotness or coldness of a substance is called temperature. It is commonly expressed in degree Celsius or Centigrade (°C) and degree Fahrenheit (°F). This climatic factor influences all plant growth processes such as photosynthesis, respiration, transpiration, breaking of seed dormancy, seed germination, protein synthesis, and translocation. At high temperatures the translocation of photosynthesis is faster so that plants tend to mature earlier.

In general, plants survive within a temperature range of 0–50 °C. The favorable or optimal day and night temperature range for plant growth and maximum yields varies among crop species.

Excessively low temperatures can also cause limiting effects on plant growth and development. For example, water absorption is inhibited when the soil temperature is low because water is more viscous at low temperatures and less mobile, and the protoplasm is less permeable. At temperatures below the freezing point of water, there is change in the form of water from liquid to solid. The expansion of water as it solidifies in living cells causes the rupture of the cell walls.

1.5.2 Relative Humidity

The amount of water vapor that the air can hold depends on its temperature. Warm air has the capacity to hold more water vapor than cold air. Relative humidity (RH) is the amount of water vapor in the air, expressed as the proportion (in percent) of the maximum amount of water vapor it can hold at certain temperature. The amount of water vapor in the air ranges from 0.01 % by volume at the frigid poles to 5 % in the humid tropics. In relation to each other, high RH means that the air is moist while air with minimal content of moisture is described as dry air. Compared to dry air, moist air has a higher RH with relatively large amounts of water vapor per unit volume of air. The RH affects the opening and closing of the stomata which regulates loss of water from the plant through transpiration as well as photosynthesis. A substantial understanding of this climatic factor is likewise important in plant propagation (Bareja 2011).

1.5.3 Wind Speed

Wind Speed and direction have significant influence on crop growth.

Beneficial impact of wind

1. Wind increases the turbulence in atmosphere, thus increasing the supply of carbon dioxide to the plants resulting in greater photosynthesis rates.
2. Wind alters the balance of hormones.
3. Wind increases the ethylene production in barley and rice.
4. Wind decreases gibberellin acid content of roots and shoots in rice.
5. Nitrogen concentration in both barley and rice increases with increase in wind speed (TNAU Agricultural Portal).

1.5.4 Sunshine Hours

There are three intensities of sunlight that suit plants: full sun, partial sun/partial shade, and shade. Plants requiring full sun will need at least six hours of direct sun daily. Partial sun/partial shade plants need 3–4 h of direct sun, and shade-loving plants will adapt to sites with less than two hours of direct sun or with filtered sun/filtered shade. Generally, the lower the light, the slower the plant grows. Hence, sunshine hours play an important role in context of the crop (HGIC1050).

1.6 Downscaling

It is a process of the development of climate data for a point or small area from regional climate information. The regional climate data may originate either from a climate model or from observations. Methodologies to model the hydrologic variables (e.g., precipitation) at smaller scale based on large-scale Global Climate Model (GCM) outputs are known as downscaling.

Downscaling, or regionalization, is the term given to the process of deriving finer resolution data (e.g., for a particular site) from coarser resolution GCM data. Most researchers feel that the horizontal resolution of most GCMs is generally too coarse to be used in impact models in its original format. A lot of useful information can be derived from GCMs without the need for downscaling, but it is recognized that sometimes it is necessary to try and add value to a scenario by making it more applicable for finer resolution studies (Barrow 2001).

Most concerns are related to the fact that regional climate is affected by forcing and circulations, which occur at sub-grid scale and hence, are not explicitly taken into account at the scales, at which GCMs operate. It may be possible to define a relationship, or relationships, between site climate and large-scale (i.e., GCM grid box scale) climate which can then be used to derive more realistic values of the future climate at the site scale. The schematic diagram is shown in Fig. 1.1.

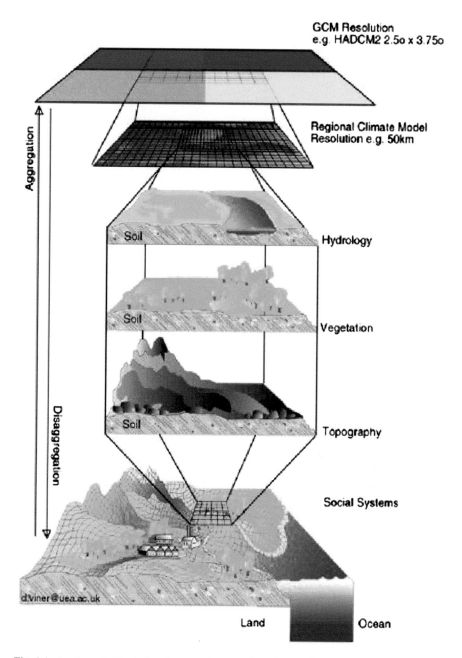

Fig. 1.1 A schematic illustrating the general approach to downscaling

Using an analogy borrowed, downscaling GCM is similar to improvements made in digital photography. Early digital photographs were not sharp but rather showed fuzzy outlines of the object in question because there was not enough detailed information. As memory systems improved and more pixels were added, the photographs became clearer and supplied a more in-depth perspective. The additional pixels needed to show accurate information that fit into the overall picture. Large-scale GCMs are a bit like early digital photography whereas downscaled GCMs or RCMs (regional climate models) provide additional information to make a clearer picture.

Downscaling methodologies fall into different categories, based on the very different approaches used, to resolve climate parameters from regional to local scales. These result in information at substantially finer resolutions (smaller areas) than GCMs provide.

There are limitations in the use of this downscaling methodology, and while they are becoming less serious, they are still important to note.

Past data must be available from several weather stations in the region of interest and those stations must be sufficiently close together to allow estimating climatic parameters over the area.

There can be no significant change in the statistical relationship between the measurements of past weather and the future projections. Another issue associated with the use of all models and their projections relates to uncertainties which can come from three sources: uncertainty around future emissions; we do not know for sure how people will behave in the future but until about 2040 all the warming scenarios are very similar because of inertia in the system and past and present emissions. Natural climate variability has some built-in chaos and will vary from decade to decade in the future as it has in the past, and modeling uncertainties, e.g., different models represent the ocean/atmosphere interactions somewhat differently.

1.6.1   Uncertainty

Confidence in global-scale GCM projections is based on well-understood physical processes and laws, the ability of GCMs to accurately simulate past climate, and the agreement in results across models. Multiple model comparisons unanimously project warming of globally averaged near-surface temperature over the next two decades in response to increased greenhouse gas emissions. However, the magnitude of this increase varies from one model to another. Additionally, in certain regions, different models project opposite changes in rainfall amount, which highlights the uncertainty of future climate change projections even when sophisticated state-of-the art GCM tools are used (Trzaska and Schnarr 2014).

There are four main sources of uncertainty in climate projections:

1. Uncertainty in future levels of anthropogenic emissions and natural forcings (e.g., volcanic eruptions).
2. Uncertainty linked to imperfect model representation of climate processes.
3. Imperfect knowledge of current climate conditions that serve as a starting point for projections.
4. Difficulty in representing interannual and decadal variability in long-term projections.

1.7 Downscaling Techniques and Their Application

Downscaling techniques have been developed, tested, and used through the efforts of many climatologists and hydrologists. More recently, downscaling has found wide application in hydroclimatology for scenario of construction, simulation, and prediction of

   (i)   regional precipitation
  (ii)   low-frequency rainfall events
 (iii)   mean, minimum, and maximum air temperature
 (iv)   soil moisture
  (v)   runoff and stream flows
 (vi)   groundwater levels
(vii)   transpiration, wind speed, and potential evaporation rates
(viii)   soil erosion and crop yield
 (ix)   landslide occurrence
  (x)   water quality.

The approaches, which have been proposed for downscaling GCMs, could be broadly classified into three categories:

Dynamical downscaling,
Statistical downscaling, and
Statistical—dynamical downscaling.

1.7.1 Dynamical Downscaling

Dynamical downscaling (DD) method involves the development of the regional climate model which required the user to highly understand the atmospheric physical behavior and local or regional interactions and feedback. Generally, DD method is used for regions of complex topography, coastal, or island locations in the regions of highly heterogeneous land cover.

The advantages cited for DD are, they respond in physically consistent ways to different external forcings, resolve the atmospheric process such as topographic precipitation and consistency with GCM. The disadvantages of DD are that it requires significant computing resources, dependent on the realism of GCM boundary forcings and initial boundary conditions.

One of the strong features of DD is that it can resolve smaller scale atmospheric features such as orographic precipitation. Complicated in design, inflexible in practical use, and high computational cost are some drawbacks associated with dynamic downscaling, and due to these drawbacks it is not highly applicable to climate change impact studies (Harun et al. 2008).

The key features of DD methods are discussed below:

What it Provides?

- 20–50 km grid cell information.
- Information at sites with no observational data.
- Daily time series.
- Scenarios for extreme events.

What it Requires?

- High computational resources and expertise.
- High volume of data inputs.
- Reliable GCM simulations.

Its Advantages

- Based on consistent, physical mechanism.
- Resolves atmospheric and surface processes occurring at sub-GCM grid scale.
- Not constrained by historical record so that novel scenarios can be simulated.
- Experiments involving an ensemble of RCMs are becoming available for uncertainty analysis.

Its Disadvantages

- Due to computational demands, RCMs are typically driven by only one or two GMC emission scenario simulations.
- Limited number of RCMs are available.
- Results depend on RCM assumptions, different RCMs will give different results.
- Affected by bias of driving GCM.

Its Applications

- Country or regional-level (e.g., European Union) assessments with significant government support and resources.
- Future planning by government agencies across multiple sectors.
- Impact studies that involve various geographic areas (Trzaska and Schnarr 2014).

1.7.2 Statistical Downscaling

Statistical downscaling or empirical downscaling is a tool for downscaling climate information from coarse spatial scales to finer scales. Statistical downscaling methods rely on empirical relationships between local-scale predictand and regional-scale predictors to downscale GCM scenarios. Successful statistical downscaling is thus dependent on long reliable series of predictors and predictand. Statistical downscaling (SD) methods are used to achieve the climate change information at the fine resolution through the development of direct statistical relationships between large-scale atmospheric circulation and local variables (such as precipitation and temperature).

Compared to other downscaling methods (e.g., dynamical downscaling), the statistical method is relatively easy to use and provides station-scale climate information from GCM-scale outputs (Wilby et al. 2002). Thus, statistical downscaling methods are the most widely used in anticipated hydrologic impact studies under climate change scenarios.

The main advantages of statistical downscaling are that they are cheap, computationally undemanding, and readily transferable. It provides the local information most needed in many climate change impact applications and ensembles of climate scenarios.

The disadvantages of statistical downscaling are that it requires highly quality data for model calibration. The predictor–predictand relationships are often nonstationary and the empirically based technique does not account for possible systematic changes in regional forcing conditions or feedback processes.

Statistical downscaling methods are particularly useful in heterogeneous environment with complex physiography or steep environment gradients (as in island, mountainous, land, and sea contexts) where there are strong relationships to synoptic scale forcing. A further justification for statistical downscaling is the need for better sub-GCM grid-scale information on extreme events such as heavy precipitation.

The real pragmatic reason is when there are severe limitations on computational resources, especially in developing nations where the greatest need exists, it has been widely recognized that statistical downscaling methods offer several practical advantages over DD procedures, especially in terms of flexible adaptation to specific study purposes and inexpensive computing resource requirements (Harun et al. 2008).

What It Provides?

- Any scale, down to station-level information.
- Daily time series (only some methods).
- Monthly time series.
- Scenarios for extreme events (only some methods).
- Scenarios for any consistently observed variable.

What it Requires?

- Medium/low computational resources.
- Medium/low volume of data inputs.
- Sufficient amount of good quality observational data.
- Reliable GCM simulations.

Its Advantages

- Computationally inexpensive and efficient, which allows for many different emission scenarios and GCM pairings.
- Methods range from simple to elaborate and are flexible enough to tailor for specific purposes.
- The same method can be applied across regions or the entire globe, which facilitates comparisons across different case studies.
- Tools are freely available and easy to implement and interpret; some methods can capture extreme events.

Its Disadvantages

- High quality observed data may be unavailable for many areas or variables.
- Assumes that relationships between large and local-scale processes will remain the same in the future (stationarity assumptions).
- The simplest methods may only provide projections at a monthly resolution.

Its Applications

- Weather generators in widespread use for crop yield, water, and other natural resource modeling and management.
- Delta or change factor method can be applied for most adaptation activities (Trzaska and Schnarr 2014).

1.7.3 Statistical–Dynamical Downscaling

Statistical–dynamical downscaling links global and regional model simulations through statistics derived for large-scale weather types. The regional simulations are initialized using representative vertical profiles for each weather type and then run for a short period without lateral forcing by the global model. The statistical–dynamical approach combines advantages of the other two methods. As in DD, a regional model is used and as in statistical–empirical downscaling, the computational effort does not depend on the length of the period to be downscaled. Statistical–dynamical downscaling consists of three steps which are described below:

I. A multiyear time series from a GCM simulation is classified into an adequate amount of large-scale weather types characteristic for the region of interest. These weather types are defined on a scale which is well resolved by the GCM.

II. Regional model simulations are carried out once for each weather type. The regional model calculates the mesoscale deviations from the large-scale state due to the impact of the regional topography. The model domain is situated within the area in which the frequencies of the large-scale weather types are derived.

III. The regional model output is weighted with the respective frequencies of the weather types and then is statistically evaluated to yield regional distributions of climatological parameters (mean values, or frequency distributions) corresponding to the global climate represented by the GCM data (Harun et al. 2008).

Statistical downscaling methodology can be broadly classified into three categories as follows:

Weather Generators
Weather generators are statistical models of observed sequences of weather variables. They are also known by stochastic weather generators. They can also be regarded as complex random number generators, the output of which resembles daily weather data at a particular location. There are two fundamental types of daily weather generators based on the approach to model daily precipitation occurrence: the Markov chain approach and the spell-length approach. In the Markov chain approach, a random process is constructed which determines a day at a station as rainy or dry, conditional upon the state of the previous day. In case of spell-length approach, instead of simulating rainfall occurrences day by day, spell-length models operate by fitting probability distribution to observed relative frequencies of wet and dry spell lengths.

Weather Typing
Weather typing approaches involve grouping of local, meteorological variables in relation to different classes of atmospheric circulation. Future regional climate scenarios are constructed either by resampling from the observed variable distribution (conditioned on the circulation pattern produced by a GCM), or by first generating synthetic sequences of weather pattern using Monte Carlo techniques and then resampling from the generated data. The mean or frequency distribution of the local climate is then derived by applying weights to the local climate states with the relative frequencies of the weather classes.

Transfer Function/Regression Method
The most popular approach of downscaling is the use of transfer function which is a regression-based downscaling method. The transfer function method relies on direct quantitative relationship between the local-scale climate variable (predictand) and the variables containing the large-scale climate information (predictors) through some form of regression. Individual downscaling schemes differ according to the

Table 1.2 Statistical downscaling category, method, predictor and predictand variables, advantages, and disadvantages

Category and Method		Predictor and Predictand	Advantages	Disadvantages
Linear methods *spatial*	*Delta method*	Same type of variable (e.g., both monthly temperature, both monthly precipitation)	• Relatively straightforward to apply • Employs full range of available predictor variables	• Requires normality of data (e.g., monthly temperature, monthly precipitation, long-term average temperature) • Cannot be applied to non-normal distributions (e.g., daily rainfall) • Not suitable for extreme events
	Simple and multiple linear regression	Variables can be of the same type or different (e.g., both monthly temperature or one monthly wind and the other monthly precipitation)		
	CCA & SVD			
Weather classification *Spatial and temporal*	*Analog method*	Variables can be of the same type or different (e.g., both monthly temperature, one large-scale atmospheric pressure field and the other daily rainfall)	• Yields physically interpretable linkages to surface climate • Versatile, i.e., can be applied to both normally and non-normally distributed data	• Requires additional step of weather-type classification • Requires large amount of data and some computational resources • Incapable of predicting new values that are outside the range of the historical data
	Cluster analysis			
	ANN			
	SOM			
Weather generator *Spatial and temporal*	*LARS-WG*	Same type of variable, different temporal scales (e.g., predictor is monthly precipitation and predictand is daily precipitation)	• Able to simulate length of wet and dry spells • Produces large number of series, which is valuable for uncertainty analysis • Production of novel scenarios	• Data-intensive • Sensitive to missing or erroneous data in the calibration set • Only some weather generators can check for the coherency between multiple variables (e.g., high insolation should not be predicted on a rainy day) • Requires generation of multiple time series and statistical postprocessing of results
	MarkSim			
	GCM			
	NHMM	Variables can be of the same type or different (e.g., both monthly temperature, one large-scale atmospheric pressure and the other daily rainfall)		

choice of mathematical transfer function, predictor variables, or statistical fitting procedure (Gautam 2010).

The statistical downscaling advantages and disadvantages of methods, predictor, and predictand variables are shown in Table 1.2.

1.8   Multiple Linear Regression

When one includes more than one predictor variable, we have what is now a multiple linear regression (MLR) model. This new model is just an extension of the simple model where we now include parameter (i.e., slope) estimates for each predictor variable in the model. These coefficient values for each predictor are the slope estimates. As with simple linear regression, we have one Y or response variable (also called the dependent variable), but now have more than one X variable, also called explanatory, independent, or predictor variables. The MLR model is as follows:

$$Y = \beta_0 + \beta_1 X1 + \cdots + \beta_k Xk + \varepsilon$$

where Y is the response variable and $X1, \ldots, Xk$ are independent variables. $\beta_0, \beta_1 \ldots \beta_k$ are fixed parameters and are random variables representing the error, or residuals, that is normally distributed with mean 0 and having a variance σ_ε^2.

1.9   Principal Component Analysis (PCA)

"PCA is a way of identifying patterns in data, and expressing the data in such a way as to highlight their similarities and differences. Since patterns in data can be hard to find in data of high dimension, where the luxury of graphical representation is not available, PCA is a powerful tool for analyzing data" (Joilliffe 2002).

Often, the variables under study are highly correlated and as such they are effectively "saying the same thing." It may be useful to transform the original set of variables to a new set of uncorrelated variables called principal components (Agrawal and Rao).

1.10   Objectives

- To develop the models for the estimation of the crop yield using MLR considering climatological parameters such as maximum temperature, minimum temperature, RH, wind speed, and sunshine hours.

- To develop the models for the estimation of the crop yield using Principal Component Regression (PCR) considering climatological parameters such as maximum temperature, minimum temperature, RH, wind speed, and sunshine hours.
- To study the MLR model developed for the estimation of crop yield by evaluating the models with the performance indices such as root mean square error (RMSE), coefficient of correlation (r), coefficient of determination (R^2), and discrepancy ratio (D.R.).
- To study the PCR model developed for the estimation of crop yield by evaluating the models with the performance indices such as root mean square error (RMSE), coefficient of correlation (r), coefficient of determination (R^2), and discrepancy ratio (D.R.).
- To compare the results of developed MLR and PCR models.

Chapter 2
Principal Component Analysis in Transfer Function

Abstract This chapter explores the transfer function in detail, with multiple linear regressions, and principal component analysis (PCA). Furthermore, it contains the slight description of various types of regression and emphasizes on the PCA and the calculations of principal components (PCs) in detail.

Keywords Transfer function · Regression methods · PCA · PCR

2.1 Transfer Function/Regression Method

The most popular approach of downscaling is the use of transfer function which is a regression-based downscaling method. The transfer function method relies on direct quantitative relationship between the local scale climate variable (predictand) and the variables containing the large scale climate information (predictors) through some form of regression. Individual downscaling schemes differ according to the choice of mathematical transfer function, predictor variables, or statistical fitting procedure. To date, linear and nonlinear regression, artificial neural network (ANN), canonical correlation, etc., have been used to derive predictor–predictand relationship. Among them, ANN-based downscaling techniques have gained wide recognition owing to their ability to capture nonlinear relationships between predictors and predictand. The main strength of transfer function downscaling is the relative ease of application. The main weakness is that the models often explain only a fraction of the observed climate variability (especially in precipitation series). Transfer methods also assume validity of the model parameters under future climate conditions. The downscaling is highly sensitive to the choice of predictor variables and statistical form. The schematic diagram of transfer function is given in Fig. 2.1.

Large-scale values of particular climate variables (predictors) will be used to predict the values of the site-specific variables (predictand). The large-scale area should roughly correspond to the size of the GCM grid box. It may be necessary to

© The Author(s) 2016
T.M.V. Suryanarayana and P.B. Mistry, *Principal Component Regression for Crop Yield Estimation*, SpringerBriefs in Applied Sciences and Technology, DOI 10.1007/978-981-10-0663-0_2

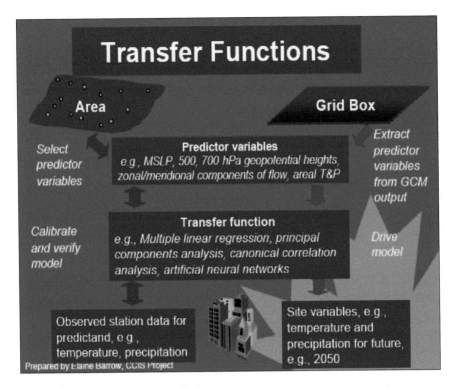

Fig. 2.1 Schematic diagram of transfer function

construct area-average values of, say, mean temperature or precipitation (usually simple averaging of station data, or weighted averaging). First step is to define the predictor variables—they must explain a high proportion of the variance in the predictand and then construct the transfer function relating the site-specific variable to the larger-scale predictors using an appropriate technique—being aware of the constraints associated with the method being used. For example, in multiple linear regressions it is assumed that the predictor variables are independent, i.e., the correlation between them is effectively zero. If this is not the case then the regression coefficients will not be a true estimate of the contribution of each of the predictor variables to the variance of the predictand. Keep back some data in order to test the performance of the model (validation).To derive the predictand values under a future climate, the larger-scale predictors derived from GCM data are used to drive the transfer function. The main advantages are firstly it is computationally much less demanding and secondly, ensembles of high resolution climate scenarios may be produced relatively easily (Barrow 2001).

2.2   Types of Regressions

2.2.1   The Simple Linear Regression Model

The relationship between a response variable Y and a predictor variable X is postulated as a linear model

$$Y = b_0 + b_1 X + E$$

where b_0 and b_1, are called the model regression coefficients, and E is a random disturbance or error. It is assumed that in the range of the observations studied, the linear equation above provides an acceptable approximation to the true relation between Y and X. In other words, Y is approximately a linear function of X, and E measures the discrepancy in that approximation.

In particular, E contains no systematic information for determining Y that is not already captured in X. The coefficient b_1, called the slope, may be interpreted as the change in Y for unit change in X. The coefficient b_0, called the constant coefficient or intercept, is the predicted value of Y, when $X = 0$.

2.2.2   The Multiple Linear Regression Model

Multiple linear regressions are with two or more independent variables on the right-hand side of the equation. Multiple linear regressions should be used, if more than one cause is associated with the effect, one wish to understand.

The equation and the true plane for the case of two independent variables, we can write the equation for a multiple regression model this way:

$$Y = \alpha + \beta X + \gamma Z + \text{error}$$

Imagine that the X- and Z-axes are on a table in front of you, with the X-axis pointing to the right and the Z-axis pointing directly away from you. The Y-axis is standing vertically, straight up from the table.

$Y = \alpha + \beta X + \gamma Z + \text{error}$ is the formula for a flat plane, that is floating in the three-dimensional space.

α is the height of the plane above the point on the table where $X = 0$ and $Z = 0$
β is the slope of the plane in the X direction, how fast the plane rises as one go to the right

If you have more than two independent variables, it is conventional to go to a subscript notation for the variables and the slope parameters as given in the equation given below.

$$Y = \alpha + \beta_1 X_1 + \beta_2 X_2 + \beta_3 X_3 + \beta_4 X_4 \ldots + \text{error}$$

2.2.3 Polynomial Regression Models

A model is said to be linear when it is linear in parameters. So the model

$$Y = \beta_0 + \beta_1 X + \beta_2 X^2 + \text{error}$$

$$Y = \alpha + \beta_1 X_1 + \beta_2 X_2 + \beta_{11} X_1^2 + \beta_{22} X_2^2 + \beta_{12} X_1 X_2 + \text{error}$$

are also the linear model. In fact, they are the second order polynomials with one and two variables, respectively.

The polynomial models can be used in those situations, where the relationship between the study and the explanatory variables is curvilinear. Sometimes, a nonlinear relationship in a small range of explanatory variables can also be modeled by the polynomials.

The Kth order polynomial model with one variable is given by the below given equation.

$$Y = \beta_0 + \beta_1 X + \beta_2 X^2 + \cdots + \beta_k X^k + \text{error}$$

2.2.4 Nonlinear Regression

In statistics, nonlinear regression is a form of analysis, in which observational data are modeled by a function, which is a nonlinear combination of the model parameters and depends on one or more independent variables. The data are fitted by a method of successive approximations.

2.3 Principal Component Analysis (PCA)

Principal Component Analysis (PCA) involves a mathematical procedure that transforms a number of possibly correlated variables into a smaller number of uncorrelated variables called PC. Principal components are also known by generation of a new set of variables by PCA, and they have a linear combination of the original variables. In this study, due to large dimensionality of predictor variables, it may be computationally unstable. Hence, PCA is performed to reduce the dimensionality of the predictor variables. PCA is also used to downscale GCM outputs of large-scale climatic variables to sub-divisional level.

> PCA is a way of identifying patterns in data, and expressing the data in such a way as to highlight their similarities and differences. Since patterns in data can be hard to find in data of high dimension, where the luxury of graphical representation is not available, PCA is a powerful tool for analyzing data.

Often, the variables under study are highly correlated and as such they are effectively "saying the same thing". It may be useful to transform the original set of variables to a new set of uncorrelated variables called PC. These new variables are linear combinations of original variables and are derived in decreasing order of importance so that the first principal component accounts for as much as possible of the variation in the original data. Also, PCA is a linear dimensionality reduction technique, which identifies orthogonal directions of maximum variance in the original data, and projects the data into a lower-dimensionality space formed of a subset of the highest variance components (Agrawal and Rao).

2.3.1 Advantages and Disadvantages of PCA

Principal component analysis (PCA) is a standard tool in modern data analysis in diverse fields from neuroscience to computer graphics—because it is a simple, nonparametric method for extracting relevant information from confusing data sets. With minimum effort, PCA provides a roadmap for how to reduce a complex data set to a lower dimension to reveal the sometimes hidden, simplified structures that often underlie it.

Another field of use is pattern recognition and image compression, therefore PCA is suited for use in facial recognition software for example, as well as for recognition and storing of other biometric data. PCA is also used in research of agriculture, biology, chemistry, climatology, demography, ecology, food research, genetics, geology, meteorology, oceanography, psychology, quality control, etc. PCA has been used in economics and finance to study changes in stock markets, commodity markets, economic growth, exchange rates, etc. Earlier studies were done in economics, but stock markets were also under research already. "Principal component or factor analysis has been used in several recent empirical studies concerned with the existence of general movements in the returns from common stocks."

PCA is a special case of factor analysis that is highly useful in the analysis of many time series and the search for patterns of movement common to several series (true factor analysis makes different assumptions about the underlying structure and solves eigenvectors of a slightly different matrix). This approach is superior to many of the bivariate statistical techniques used earlier, in that it explores the interrelationships among a set of variables caused by common "factors," mostly economic in nature. PCA is a way of identifying patterns in data, and expressing the data in such a way as to highlight their similarities and differences. A primary benefit of PCA arises from quantifying the importance of each dimension for describing the variability of a data set. PCA can also be used to compress the data, by reducing the number of dimensions, without much loss of information. When using PCA to analyze a data set, it is usually possible to explain a large percentage of the total variance with only a few components. Principal components are selected so that each successive one explains a maximum of the remaining variance; the first component is selected to explain the maximum proportion of the total variance, the second to explain the maximum of the remaining variance, etc. PCA is completely nonparametric: any data set can be plugged in and an answer comes out, requiring no parameters to tweak and no regard for how the data was recorded. From one perspective, the fact that PCA is nonparametric (or plug-and-play) can be considered a positive feature because the answer is unique and independent of the user (Kumar and chauhan 2014).

2.3.2 Applications of Principal Components

The most important use of PCA is reduction of data. It provides the effective dimensionality of the data. If first few components account for most of the variation in the original data, then first few components' scores can be utilized in subsequent analysis in place of original variables.

Plotting of data becomes difficult with more than three variables. Through PCA, it is often possible to account for most of the variability in the data by first two components, and it is possible to plot the values of first two components scores for each individual. Thus, PCA enables us to plot the data in two dimensions. Particularly, detection of outliers or clustering of individuals will be easier through this technique. Often, use of PCA reveals grouping of variables, which would not be found by other means.

Reduction in dimensionality can also help in analysis, where number of variables is more than the number of observations, for example, in discriminant analysis and regression analysis. In such cases, PCA is helpful by reducing the dimensionality of data.

Multiple regressions can be dangerous, if independent variables are highly correlated. PCA is the most practical technique to solve the problem. Regression analysis can be carried out using PC as regressors in place of original variables. This is known as principal component regression (Agrawal and Rao).

2.4   Principal Component Regression (PCR)

In order to conduct principal component regression, PCA is carried out as follows:

Principal Component Analysis
When starting a research, students as well as researchers, often collect a lot of data or sometimes come across large datasets that are available. But when having lots of data, especially when it is secondary data, it is often very easy to get confused. It is hard to find the variables that are really important for the research when there are so many variables to consider. This is where principal components analysis (PCA) can help.

Principal components analysis (PCA) was invented by Karl Pearson in 1901 and is now used in many fields of science. PCA is mostly used as a tool in exploratory data analysis because what it essentially does is to find the most important variables (a combination of them) that explain most of the variance in the data. So, when there is lots of data to be analyzed, PCA can make the task a lot easier. PCA also helps to construct predictive models (Chen et al. 2011).

2.4.1   Calculating Principal Components

The Principal components (PCs) can be found using purely mathematical arguments—they are given by an orthogonal linear transformation of a set of variables optimizing a certain algebraic criterion. An overview how to perform PCA is given hereafter.

Let $x_1, x_2, x_3 \ldots x_p$ are variables under study, and then first principal component may be defined as

$$z_1 = a_{11}x_1 + a_{12}x_2 + \cdots + a_{1p}x_p$$

Such that variance of z_1 is as large as possible subject to the condition that

$$a_{11}^2 + a_{12}^2 + \cdots a_{1p}^2 = 1$$

This constraint is introduced because if this is not done, then Var (z_1) can be increased simply by multiplying any a_{1js} by a constant factor. The second principal component is defined as

$$z_2 = a_{21}x_1 + a_{22}x_2 + \cdots + a_{2p}x_p$$

Such that Var (z_2) is as large as possible next to Var (z_1) subject to the constraint that

$$a_{21}^2 + a_{22}^2 + \cdots + a_{2p}^2 = 1 \text{ and } \quad \text{cov}(z_1, z_2) = 0 \text{ and } \quad \text{so on.}$$

It is quite likely that first few PC accounts for most of the variability in the original data. If so, these few PC can then replace the initial p variables in subsequent analysis, thus reducing the effective dimensionality of the problem. An analysis of PC often reveals relationships that were not previously suspected and thereby allows interpretation that would not ordinarily result. However, PCA is more of a means to an end rather than an end in itself because this frequently serves as intermediate steps in much larger investigations by reducing the dimensionality of the problem and providing easier interpretation. It is a mathematical technique, which does not require user to specify the statistical model or assumption about distribution of original variants. It may also be mentioned that, PCs are artificial variables and often, it is not possible to assign physical meaning to them. Further, since PCA transforms original set of variables to the new set of uncorrelated variables, it is worth stressing that, if original variables are uncorrelated, and then there is no point in carrying out PCA (Agrawal and Rao).

2.4.2 Rules for Retaining Principal Components

In the initial extraction process, PCA will derive as many components as the number of measured variables. After the initial components are extracted, the analyst must decide on how many components should be retained to meaningfully represent the original correlation matrix. The initial component eigenvalues, percent of variance accounted for, and cumulative variance accounted for are provided. According to Stevens, "Probably the most widely used criterion is that of: Retain only those components whose eigenvalues are greater than 1".

A fairly common technique noted in the literature combines the two approaches. Eigen values greater than one is initially retained and the screen test is used subsequently to assess the tenability of the model. Because Eigen values represent reproduced variance, this is equivalent to setting a minimum level of acceptable variance reproduced by a component. The second stage evaluates the parsimony of the solution relative to the contribution of each component to reproducing the original variance in the data. A potential disadvantage of this approach is the arbitrary criterion of retaining eigenvalues greater than one in the first stage. Because PCA studies typically rely on sample data, eigenvalues (reproduced variance) should be expected to change (even with large samples) slightly from sample to sample. In addition, the interpretation of what constitutes a "meaningful" amount of variance accounted for (which eigenvalues represent) is inherently subjective (Kellow 2006).

2.4.3   *Development of Principal Component Regression (PCR)*

Principal component regression (PCR) is a type of regression analysis, which considers PC as independent variables, instead of adopting original variables. The PCs are the linear combination of the original variables which can be obtained by PCA. The PCA transforms the original set of intercorrelated independent variables to a new set of uncorrelated variables (i.e., PCs). The use of these PCs as independent variables is quite useful in the multiple regression models to avoid the multicollinearity problem and to identify the variables which are the most significant in making the prediction. The PCR models have been developed using PCs as inputs to predict and to compare the same with multiple linear regression models. It has been found that that the incorporation of PCs as independent variables in the regression models improved the model prediction as well as reduced the model complexity by eliminating multicollinearity.

Principal components regression (PCR) is a method for combating multicollinearity and results in estimation and prediction better than ordinary least squares, when used successfully. With this method, the original k climatic variables are transformed into a new set of orthogonal or uncorrelated variables called PC of the correlation matrix. This transformation ranks the new orthogonal variables in order of their importance and the procedure, and then involves eliminating some of the PC to effect a reduction in variance (Fekedulegn et al. 2002).

Chapter 3
Review of Literature

Abstract This chapter covers the various research works carried out in the field of analyzing climatic variability induced due to climate change scenarios and its impact on agriculture. The review of works is classified into four domains, viz. climate change, downscaling techniques, multiple linear regression, and principal component analysis and regression.

Keywords Review · Climate change · Downscaling techniques · MLR · PCA · PCR

3.1 Review of Works on Climate Change

The predictions of climate change under various emission scenarios were highly uncertain but it was expected to affect agricultural crop production in the twenty-first century. However, we know very little about future changes in specific cropping systems under climate change in California's Central Valley. Lee and Six (2010) used DAYCENT to simulate changes in yield and fluxes of greenhouse gases under A2 (medium-high) and B1 (low) emission scenarios. In total, 18 climate change predictions for the two scenarios were considered by applying different climate models and downscaling methods. The following crops were selected: alfalfa (hay), cotton, maize, winter wheat, tomato, rice, and sunflower. The simulations suggest that future climate change under the different emission scenarios will lead to a broad range of impacts on crop yields. By 2007, yields under A2 decreased in comparison to the 2009 baseline in the following order: cotton (29 %) > sunflower (27 %) > wheat (17 %) > rice (12 %) > tomato (9 %) > maize (8 %). Yields were between 5 (alfalfa) and 21 % (cotton) lower under A2 compared to B1. Under A2, soil carbon (C) storage tended to decrease under climate change due to a decrease in C inputs to the soil and an increase in soil C decomposition. However, differences in nitrous oxide (N_2O) flux between A2 and B1 were not clear.

Global warming, climate change, and tourism of late have taken the center stage of academic research. A raging debate was on, apart from the popular writings and research articles published on the theme. According to the Intergovernmental Panel

© The Author(s) 2016
T.M.V. Suryanarayana and P.B. Mistry, *Principal Component Regression for Crop Yield Estimation*, SpringerBriefs in Applied Sciences and Technology, DOI 10.1007/978-981-10-0663-0_3

on Climate Change "Warming of the climate system is unequivocal as is now evident from observations of increases in global average air and ocean temperatures, widespread melting of snow and ice since the mid-20th century." The conceptual paper by Ramasamy and Swamy (2012) is carried out by the contributions of 30 selected papers published in tourism-related journals. The approaches of this manuscript are conceptual and are self-oriented to bring readers an all-encompassed state of the art. The purpose of this study was to identify and understand the extent of research carried out to assess the impact of global warming and climate change on tourism. A three-pronged approach was adopted to collect data. First, a literature search was conducted on Google search engine; second, referred research journals in the areas of global warming, climate change, and tourism are consulted, and third, published reports of national and international scientific organizations and government organizations are examined. The fortunes of tourism industry, given the nature of activity, obviously depend on the magnitude and impact of global warming and climate change. Countries like USA, China, Russia, India, and Australia are largely attributed for the growing pollution and the consequent changes in the global climate. Sector- wise, aviation accounts for 40 %, automobiles 32 %, accommodations 21 %, and others 7 % are found to be the major contributors. Incidentally, all these sectors are related both directly and indirectly to the tourism industry.

Kumar and Sharma (2013) analysed the impact of climate change on agricultural productivity in quantity terms, value of production in monetary terms, and food security in India. The study undertook state-wise analysis based on secondary data for the duration of 1980–2009. Climate variation affects food grain and non-food grain productivity and both these factors along with other socioeconomic and government policy variables affect food security. Food security and poverty are interlinked with each other as cause and effect and vice versa, particularly, for a largely agrarian economy of India. Regression results for models proposed in this study show that for most of the food grain crops, non-food grain crops in quantity produced per unit of land, and in terms of value of production, climate variation causes negative impact. The adverse impact of climate change on the value of agricultural production and food grains indicates food security threat to small and marginal farming households. The state-wise food security index was also generated in this study; and econometric model estimation reveals that the food security index itself also gets adversely affected due to climatic fluctuations.

3.2 Review of Works on Downscaling Techniques

The impact of global warming on the temperature regime of a single site is explored by Trigo and Palutikof (1999) with reference to Coimbra in Portugal. The basis of the analysis is information taken from a climate change simulation performed with a state-of-the-art general circulation model (the Hadley Centre model). First, it is shown that the model is unable to reproduce accurately the statistics of daily maximum and minimum temperature at the site. Second, using a reanalysis data set, downscaling models are developed to predict site temperature from large-scale free atmosphere variables derived from the sea-level pressure and 500 hPa geopotential height fields. In particular, the relative performances of linear models and nonlinear artificial neural networks are compared using a set of rigorous validation techniques. It is shown that even a simple configuration of a two-layer nonlinear neural network significantly improves the performance of a linear model. Finally, the nonlinear neural network model is initialized with general circulation model output to construct scenarios of daily temperature at the present day (1970–1979) and for the future decade (2090–2099). These scenarios are analyzed with special attention to the comparison of the frequencies of heat waves (days with maximum temperature greater than 35 °C) and cold spells (days with minimum temperature below 5 °C).

Schoof and Pryor (2001) had carried out study with comparison of two statistical downscaling methods for daily maximum and minimum surface air temperature, total daily precipitation, and total monthly precipitation at Indianapolis, in USA. The analysis is conducted for two seasons, the growing season and the nongrowing season, defined based on variability of surface air temperature. The predictors used in the downscaling are indices of the synoptic scale circulation derived from rotated principal components analysis (PCA) and cluster analysis of variables extracted from an 18-year record from seven rawinsonde stations in the Midwest region of the United States. PCA yielded seven significant components for the growing season and five significant components for the nongrowing season. These PCs explained 86 and 83 % of the original rawinsonde data for the growing and nongrowing seasons, respectively. Cluster analysis of the PC scores using the average linkage method resulted in 8 growing season synoptic types and 12 non-growing synoptic types. The downscaling of temperature and precipitation are conducted using PC scores and cluster frequencies in regression models and artificial neural networks (ANNs). Regression models and ANNs yielded similar results, but the data for each regression model violated at least one of the assumptions of regression analysis. As expected, the accuracy of the downscaling models for temperature was superior to that for precipitation. The accuracy of all temperature models was improved by adding an autoregressive term, which also changed the relative importance of the dominant anomaly patterns as manifest in the PC scores. Application of the transfer functions to model daily maximum and minimum temperature data from an independent time series resulted in correlation coefficients of 0.34–0.89. In accord with previous studies, the precipitation models exhibited lesser predictive capabilities. The correlation coefficient for predicted versus observed daily precipitation totals was less than 0.5 for both seasons, while that for monthly total precipitation was below 0.65. The downscaling techniques are discussed in terms of model performance, comparison of techniques, and possible model improvements.

Researchers are aware of certain types of problems that arise when modeling interconnections between general circulation and regional processes, such as prediction of regional, local-scale climate variables from large-scale processes, e.g., by means of general circulation model (GCM) outputs. A statistical downscaling approach to monthly total precipitation over Turkey, which is an integral part of

system identification for analysis of local-scale climate variables, is investigated by
Tatli et al. (2004). Based on perfect prognosis, a new computationally effective
working method is introduced by the proper predictors selected from the National
Centers for Environmental Prediction–National Center for Atmospheric Research
reanalysis data sets, which are simulated as perfectly as possible by GCMs during
the period of 1961–1998. The Sampson correlation ratio is used to determine the
relationships between the monthly total precipitation series and the set of
large-scale processes (namely 500 hPa geopotential heights, 700 hPa geopotential
heights, sea-level pressures, 500 hPa vertical pressure velocities, and 500–1000 hPa
geopotential thicknesses). In the study, statistical preprocessing is implemented by
independent component analysis rather than principal component analysis or
principal factor analysis. The proposed downscaling method originates from a
recurrent neural network model of Jordan that uses not only large-scale predictors,
but also the previous states of the relevant local-scale variables. Finally, some
possible improvements and suggestions for further study are mentioned.

Monthly mean temperatures at 562 stations in China are estimated using a
statistical downscaling technique. The technique used by Li-Jun (2009) was multiple linear regressions (MLRs) of principal components (PCs). A stepwise
screening procedure is used for selecting the skilful PCs as predictors used in the
regression equation. The predictors include temperature at 850 hPa (T), the combination of sea-level pressure and temperature at 850 hPa (P + T), and the combination of geopotential height and temperature at 850 hPa (H + T). The
downscaling procedure is tested with the three predictors over three predictor
domains. The optimum statistical model is obtained for each station and month by
finding the predictor and predictor domain corresponding to the highest correlation.
Finally, the optimum statistical downscaling models are applied to the Hadley
Centre Coupled Model, version 3 (HadCM3) outputs under the Special Report on
Emission Scenarios (SRES) A2 and B2 scenarios to construct local future temperature change scenarios for each station and month. The results show that
(1) statistical downscaling produces less warming than the HadCM3 output itself;
(2) the downscaled annual cycles of temperature differ from the HadCM3 output,
but are similar to the observation; (3) the downscaled temperature scenarios show
more warming in the north than in the south; and (4) the downscaled temperature
scenarios vary with emission scenarios, and the A2 scenario produces more
warming than the B2, especially in the north of China.

Ojha et al. (2010) studied downscaling models using a linear multiple regression
(LMR) and artificial neural networks (ANNs) for obtaining projections of mean
monthly precipitation to lake-basin scale in an arid region in India. The effective-
ness of these techniques was demonstrated through application to downscale the
predict and (precipitation) for the Pichola lake region in Rajasthan state in India,
which was considered to be a climatically sensitive region. The predictor variables
are extracted from (1) the National Centers for Environmental Prediction (NCEP)
reanalysis dataset for the period 1948–2000, and (2) the simulations from the
third-generation Canadian Coupled Global Climate Model (CGCM3) for emission
scenarios A1B, A2, B1, and COMMIT for the period 2001–2100. The scatter plots

and cross correlations were used for verifying the reliability of the simulation of the predictor variables by the CGCM3. The performance of the linear multiple regression and ANN models was evaluated based on several statistical performance indicators. The ANN-based models are found to be superior to LMR-based models and subsequently, the ANN-based model was applied to obtain future climate projections of the predict and (i.e., precipitation). The precipitation is projected to increase in future for A2 and A1B scenarios, whereas it is least for B1 and COMMIT scenarios using predictors. In the COMMIT scenario, the emissions are held the same as in the year 2000.

An extensive statistical 'downscaling' study is done to relate large-scale climate information from a general circulation model (GCM) to local-scale river flows in SW France for 51 gaging stations ranging from nival (snow-dominated) to pluvial (rainfall-dominated) river systems. Tisseuil et al. (2010) studied to select the appropriate statistical method at a given spatial and temporal scale to downscale hydrology for future climate change impact assessment of hydrological resources. The four proposed statistical downscaling models use large-scale predictors (derived from climate model outputs or reanalysis data) that characterize precipitation and evaporation processes in the hydrological cycle to estimate summary flow statistics. The four statistical models used are generalized linear (GLM) and additive (GAM) models, aggregated boosted trees (ABT), and multi-layer perceptron neural networks (ANN). These four models were each applied at two different spatial scales, namely at that of a single flow gaging station (local downscaling) and that of a group of flow gaging stations having the same hydrological behavior (regional downscaling). For each statistical model and each spatial resolution, three temporal resolutions were considered, namely the daily mean flows, the summary statistics of fortnightly flows, and a daily "integrated approach." The results show that flow sensitivity to atmospheric factors is significantly different between nival and pluvial hydrological systems which are mainly influenced, respectively, by shortwave solar radiations and atmospheric temperature. The non-linear models (i.e., GAM, ABT, and ANN) performed better than the linear GLM when simulating fortnightly flow percentiles. The aggregated boosted trees method showed higher and less variable R^2 values to downscale the hydrological variability in both nival and pluvial regimes. Based on GCM cnrm-cm3 and scenarios A2 and A1B, future relative changes of fortnightly median flows were projected based on the regional downscaling approach. The results suggest a global decrease of flow in both pluvial and nival regimes, especially in spring, summer, and autumn, whatever the considered scenario. The discussion considers the performance of each statistical method for downscaling flow at different spatial and temporal scales as well as the relationship between atmospheric processes and flow variability.

Aksornsingchai and Srinilta (2011) studied three statistical downscaling methods to predict temperature and rainfall at 45 weather stations in Thailand. Methods under consideration are multiple linear regressions (MLR), support vector machine with polynomial kernel (SVM-POL), and support vector machine with radial basis function kernel (SVM-RBF). Large-scale data are from Geophysical Fluid Dynamics Laboratory (GFDL). Five predictor variables are chosen: (1) temperature,

(2) pressure, (3) precipitation, (4) evaporator, and (5) net short wave. Accuracy is assessed by tenfold cross-validation in terms of root-mean-squared error (RMSE) and correlation coefficient (R). SVM-RBF is the most accurate model. Prediction accuracy of monthly average rainfall and temperature is satisfying in most part of the country. Lastly, downscaling models can project long-term trends of monthly average rainfall and temperature.

The summer rainfall over the middle-lower reaches of the Yangtze River valley (YRSR) has been estimated by Yan et al. (2011) with a multi-linear regression model using principal atmospheric modes derived from a 500 hPa geopotential height and a 700 hPa zonal vapor flux over the domain of East Asia and the West Pacific. The model was developed using data from 1958 to 1992 and validated with an independent prediction from 1993 to 2008. The independent prediction was efficient in predicting the YRSR with a correlation coefficient of 0.72 and a relative root-mean-square error of 18 %. The downscaling model was applied to two general circulation models (GCMs) of Flexible Global Ocean-Atmosphere-Land System Model (FGOALS) and Geophysical Fluid Dynamics Laboratory coupled climate model version 2.1 (GFDL-CM2.1) to project rainfall for present and future climate under B1 and A1B emission scenarios. The downscaled results provided a closer representation of the observation compared to the raw models in the present climate. In addition, compared to the inconsistent prediction directly from different GCMs, the downscaled results provided a consistent projection for this half-century, which indicated a clear increase in the YRSR. Under the B1 emission scenario, the rainfall could increase by an average of 11.9 % until 2011–2025 and 17.2 % until 2036–2050 from the current state; under the A1B emission scenario, rainfall could increase by an average of 15.5 % until 2011–2025 and 25.3 % until 2036–2050 from the current state. Moreover, the increased rate was faster in the following decade (2011–2025) than the latter of this half-century (2036–2050) under both emissions.

Downscaling is a technique for obtaining local-scale hydrological variables from coarser-scale atmospheric variables that are generated by general circulation models. Mainly there are two downscaling methods, i.e., dynamic downscaling and statistical downscaling. Statistical downscaling offers less computational work compared to dynamic downscaling and it also provides a platform to use ensemble GCM outputs. Devak and Dhanya (2014), in their paper, compared the results generated from two methods, i.e., by support vector machine (SVM) and k-Nearest Neighbor (KNN), which covers some parts of Chhattisgarh, Orissa, Bihar, and Maharashtra state. The above two models are applied at five different locations in Mahanadi Basin. Bias correction by equidistant CDF matching method is also applied to the future projection. Calibration and validation of the model incorporates the result from Canadian global climate model (CanCM4) for historical scenario and future projections are done using predictors from RCP 4.5 scenario. Various performance measures like, normalized mean square error (NMSE) and correlation coefficient is also taken into account. Kolmogorov Smirnov test is also performed for the two models.

3.3 Review of Works on Multiple Linear Regressions

Meteorological data mining is a form of data mining concerned with finding hidden patterns inside largely available meteorological data, so that the information retrieved can be transformed into usable knowledge. Weather is one of the meteorological data that is rich in important knowledge. The most important climatic element which impacts on agricultural sector is rainfall. Thus, rainfall prediction becomes an important issue in agricultural country like India. Dutta and Tahbilder (2014) used data mining technique in forecasting monthly rainfall of Assam. This was carried out using traditional statistical technique—multiple linear regression. The data include six-year period (2007–2012) collected locally from Regional Meteorological Center, Guwahati, Assam, India. The performance of this model is measured in adjusted R-squared. Our experiment results show that the prediction model based on multiple linear regression indicates acceptable accuracy.

Agrarian sector in India is facing rigorous problem to maximize the crop productivity. More than 60 % of the crop still depends on monsoon rainfall. Recent developments in information technology for agriculture field has become an interesting research area to predict the crop yield. The problem of yield prediction is a major problem that remains to be solved based on available data. Data mining techniques are the better choices for this purpose. Different data mining techniques are used and evaluated in agriculture for estimating the future year's crop production. Ramesh and Vardhan (2015) presented a brief analysis of crop yield prediction using multiple linear regression (MLR) technique and density-clustering technique for the selected region, i.e., East Godavari district of Andhra Pradesh in India.

3.4 Review of Works on Principal Component Analysis and Principal Component Regression

Tenderness is the most important factor affecting consumer prediction of eating quality of meat. Park et al. (2001) developed the principal component regression (PCR) models to relate near-infrared (NIR) reflectance spectra of raw meat to Warner–Bratzler (WB) shear force measurement of cooked meat. NIR reflectance spectra with wavelengths from 1100 to 2498 nm were collected on 119 longissimus dorsi meat cuts. The first principal component (or factor) from the absorption spectra $\log(1/R)$ showed that the most significant variance from the spectra of tough and tender meats were due to the absorptions of fat at 1212, 1722, and 2306 nm and water at 1910 nm. The distinctive fat absorption peaks at 1212, 1722, 1760, and 2306 nm were found in the second factor of the second derivative spectra of meat. In addition, the local minima in the second principal component of the second derivative spectra showed the importance of water absorption at 1153 nm and protein absorption at 1240, 1385, and 1690 nm. When the absorption spectra between 1100 and 2498 nm were used, the coefficient of determination (R^2) of the PCR model to predict WB shear force tenderness was 0.692. The R^2 was 0.612 when the spectra between 1100 and 1350 nm were analysed. When the second derivatives of the spectral data were used, the R^2 of the PCR model to predict WB shear force of the meat was 0.633 for the full spectral range of 1100–2498 nm and 0.616 for the spectral range of 1100–1350 nm.

Webster (2001) studied principal component regression analysis to examine the relative contributions of 11 ranking criteria used to construct the U.S. News & World Report (USNWR) tier rankings of national universities. The main finding of the study was that the actual contributions of the 11 ranking criteria examined difference substantially from the explicit USNWR weighting scheme because of severe and pervasive multicollinearity among the ranking criteria. USNWR assigns the greatest weight to academic reputation. However, generated first principal component eigenvalues of tier rankings indicate that the most significant ranking criterion was the average SAT scores of enrolled students. This result was significant since admission requirements are policy variables that indirectly affect, for example, admission applications, yields, enrollment, retention, tuition-based revenues, and alumni contributions.

Principal component analysis is one of the most widely applied tools in order to summarize common patterns of variation among variables. Several studies have investigated the ability of individual methods, or compared the performance of a number of methods, in determining the number of components describing common variance of simulated data sets. Peres Neto et al. (2005) identified a number of shortcomings related to these studies and conducted an extensive simulation study where they compared a larger number of rules available and developed some new methods. In total, we compare 20 stopping rules and propose a two-step approach that appears to be highly effective. First, a Bartlett's test is used to test the significance of the first principal component, indicating whether or not at least two variables share common variation in the entire data set. If significant, a number of different rules can be applied to estimate the number of non-trivial components to be retained. However, the relative merits of these methods depend on whether data contain strongly correlated or uncorrelated variables. Also estimated the number of non-trivial components for a number of field data sets so that, one can evaluate the applicability of our conclusions based on simulated data.

As a useful alternative to the Cox proportional hazards model, the linear regression survival model assumes a linear relationship between the covariates and a known monotone transformation, for example, logarithm of an event time of interest. Ma (2007), in their article, studied the linear regression survival model with right censored survival data, when high-dimensional microarray measurements are present. Such data may arise in studies investigating the statistical influence of molecular features on survival risk. They proposed using the principal component regression (PCR) technique for model reduction based on the weight least squared Stute estimate. Compared with other model reduction techniques, the PCR approach was relatively insensitive to the number of covariates and hence suitable for high-dimensional microarray data. Component selection based on the nonparametric bootstrap, and model evaluation using the time-dependent ROC (receiver operating characteristic) technique are investigated. They demonstrated the proposed approach with datasets from two microarray gene expression profiling studies of lymphoma cancers.

The main purpose of this study by Mendes (2011) was to show that how one can use multivariate multiple linear regression analysis (MMLR) based on principal component scores to investigate relations between two data sets (i.e., pre- and posts laughter traits of Ross 308 broiler chickens). Principal component analysis (PCA) was applied to predictor variables to avoid multicollinearity problem. According to results of the PCA, out of 7 principal components, only the first three components (PC1, PC2, and PC3) with eigenvalue greater than 1 were selected (explained 89.45 % of the variation) for MMLR analysis. Then, the first three prin-cipal component scores were used as predictor variables in MMLR. The results of MMLR analysis showed that shank width, breast circumference, and body weight had a similar linear effect on predicting the post-slaughter traits (P = 0.746). As a result, since the animals had high value of shank width, breast circumference, and body weight, it might be probable that their post-slaughter traits namely heart weight, liver weight, gizzard weight, and hot carcass weight were also expected to be high.

Principal component analysis (PCA) and multiple linear regressions were applied on the surface water quality data by Mustapha and Abdu (2012) with the aim of identifying the pollution sources and their contribution toward water quality varia-tion. Surface water samples were collected from four different sampling points along Jakara River. Fifteen physico-chemical water quality parameters were selected for analysis: dissolved oxygen (DO), biochemical oxygen demand (BOD5), chemical oxygen demand (COD), suspended solids (SS), pH, conductivity, salinity, temper-ature, nitrogen in the form of ammonia (NH_3), turbidity, dissolved solids (DS), total solids (TS), nitrates (NO_3), chloride (Cl), and phosphates (PO_4^3). PCA was used to investigate the origin of each water quality parameters and yielded five varimax factors with 83.1 % total variance, and in addition PCA identified five latent pol-lution sources namely: ionic, erosion, domestic, dilution effect, and agricultural run-off. Multiple linear regressions identified the contribution of each variable with significant values.

In recent decades, particulate matter is one of the prevalent pollutants recorded throughout Malaysia. The development of models to predict particulate matter less than and equal 10 μm (PM_{10}) concentration is thus very useful because it can provide early warning to the population and for input into decision regarding abatement measures and air quality management. The aim of the study by Ul-Saufie et al. (2011) was to improve the predictive power of multiple linear regression models using principal components as input for predicting PM_{10} concentration for the next day. The developed model was compared with multiple linear regression models. Performance indicator such as prediction accuracy (PA), coefficient of determination (R^2), index of agreement (IA), normalized absolute error (NAE) and root-mean-square rrror (RMSE) were used to measure the accuracy of the models. Results showed that the use of principal component as inputs improved multiple linear regression models prediction by reducing their complexity and eliminating data collinearity.

Kelechi (2012), in their paper, used the regression analysis and principal component analysis (PCA) to examine the possibility of using few explanatory variables to explain the variation in the dependent variable. It applied regression analysis and principal component analysis (PCA) to assess the yield of turmeric, from National Root Crop Research Institute Umudike in Abia State, Nigeria. This was done by estimating the coefficients of the explanatory variables in the analysis. The explanatory variables involved in this analysis show a multiple relationship between them and the dependent variable. A correlation table was obtained from which the characteristic roots were extracted. Also, the orthonormal basis was used to establish the linearly independent relationships of the variables. The regression analysis shows in details the constant and the coefficients of the three explanatory variables. On the other hand, the principal component analysis estimates the first principal component and second principal component, and both components accounted for 71.4 % of the total variation. The regression analysis and principal component analysis (PCA) yielded good estimates, which lead to the structural coefficient of the regression model. The study shows that regression analysis and principal component analysis (PCA) use few explanatory variables to explain variations in a dependent variable and are therefore efficient tools for assessing turmeric yield depending on the set objective. But that PCA is more efficient, since it uses fewer variables to achieve the same result.

Accurate forecast of water demand is very crucial in developing a water resource management strategy to check the balance of future water supply and demand to ensure proper water supplies to the people. In order to forecast water demand, different models have been adopted in the literature. Among these the multiple regression analysis was quite popular for long term water demand forecasting. In spite of their evident success in modeling water demands, it can face difficulties in the case of multicollinearity, which implies highly correlated variables. Since water demand depends on many factors such as population, household size, rainfall, temperature, age of population, education, water price, and policy, a multi-collinearity problem may arise during the development of a multiple regression model which may lead to the incorrect estimation of future water demand. To avoid multicollinearity problem, principal component regression analysis has been used in several environmental studies which demonstrated its ability to eliminate the multicollinearity problem and to produce better model results. However, application of principal component regression in water demand forecasting is limited. In their study, Haque et al. (2013) developed principal component regression model by combining multiple linear regression and principal component analysis to forecast future water demand in the Blue Mountains Water Supply systems in New South Wales, Australia. In addition, performances of the developed principal component regression model were compared with multiple linear regression models by adopting several model evaluation statistics such as relative error, bias, Nash-Sutcliffe efficiency, and accuracy factor. It was found that the developed principal component regression model was able to predict future water demand with a higher degree of accuracy with an average relative error, bias, Nash-Sutcliffe efficiency, and accuracy factor values of 3.4, 2.92, 0.44, and 1.04 %, respectively. Moreover, it was found that the principal component regression model performed better than the multiple linear regression models and could be used to eliminate the multi-collinearity problem. The method presented in their paper can be adapted to other cities in Australia and the world.

Application of principal component analysis in developing statistical models for forecasting crop yield has been demonstrated. The time series data on wheat yield and weekly weather variables, viz., minimum and maximum temperature, relative humidity, wind- velocity and sunshine hours pertaining to the period 1990–2010 in Faizabad district of Uttar Pradesh have been used in this study. Weather indices have been constructed using weekly data on weather variables. Yadav et al. (2014) developed four models using principal component analysis as regressor variables including time trend and wheat yield as regressand. The model 1 and 3 have been found to be most appropriate on the basis of R^2adj, percent deviation of forecast, RMSE (%), and PSE for the forecast of wheat yield 2 months before the harvest of the crop.

Chapter 4
Study Area and Data Collection

Abstract This chapter contains the brief overview of agroclimatic zones of India followed by the subagroclimatic zones of Gujarat. Thereafter, the slight description of the study area and the data required for the study are provided.

Keywords Agroclimatic zones · Subagroclimatic zones

4.1 Agroclimatic Zones by the Planning Commission

The Planning Commission, as a result of the midterm appraisal of the planning targets of the Seventh Plan, has divided the country into 15 broad agroclimatic zones based on physiography, soils, geological formation, climate, cropping patterns, and development of irrigation and mineral resources for broad agricultural planning and developing future strategies. Fourteen zones were in the main land and the remaining one in the islands of Bay of Bengal and the Arabian Sea. The agroclimatic zones of India are illustrated in Fig. 4.1.

4.2 Subagroclimatic Zones of Gujarat

Gujarat lies in the agroclimatic zone-XIII, which is called as "Gujarat Plains and Hills region". Traditionally Gujarat was divided into three regions, namely, (i) the mainland plains extending from the Rann of Kutch and the Aravalli Hills in the north to Damanganga in the south, (ii) the hilly peninsular region of Saurashtra and the rocky areas of Kutch and (iii) the northeastern hill tract. Now it is divided into seven subagroclimatic zones: Southern Hills (Dangs, Valsad), Southern Gujarat, Middle Gujarat, North Gujarat, Northwest Arid, North Saurashtra and South Saurashtra. Most of Gujarat falls under mega thermic category with mean soil

© The Author(s) 2016
T.M.V. Suryanarayana and P.B. Mistry, *Principal Component Regression for Crop Yield Estimation*, SpringerBriefs in Applied Sciences and Technology,
DOI 10.1007/978-981-10-0663-0_4

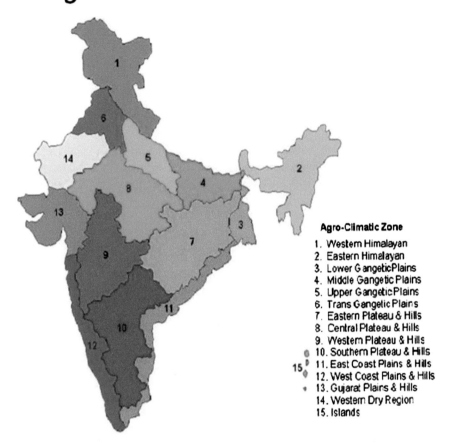

Agro-Climatic Zone
1. Western Himalayan
2. Eastern Himalayan
3. Lower Gangetic Plains
4. Middle Gangetic Plains
5. Upper Gangetic Plains
6. Trans Gangetic Plains
7. Eastern Plateau & Hills
8. Central Plateau & Hills
9. Western Plateau & Hills
10. Southern Plateau & Hills
11. East Coast Plains & Hills
12. West Coast Plains & Hills
13. Gujarat Plains & Hills
14. Western Dry Region
15. Islands

Fig. 4.1 Agroclimatic zones of India

temperature exceeding 28 °C. Air temperature in January normally remains over 10 °C. Maximum temperature in May goes over 40 °C in north and Northwest Gujarat. It is more moderate in the coastal area of Saurashtra and South Gujarat. Rainfall is the most dominant climatic factor. Average rainfall is 828 mm which is received in 35 days mostly from June to September with a coefficient of variation (CV) of 50 %. Spatially it ranges from 300 mm in northwest to 2000 mm in southeast. 12 out of 26 districts of Gujarat are drought prone. In years of poor rainfall, the yields of important crops like groundnut which are mainly rainfed can reduce by 70 % or more. The details of the subagroclimatic zones are given in Fig. 4.2 and discussed thereafter.

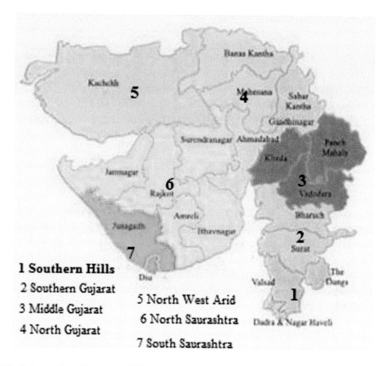

Fig. 4.2 Subagroclimatic zones of Gujarat

4.2.1 *Southern Hills*

This is small but largely a tribal belt covering the districts of the Dangs and Valsad. The area receives about 1793 mm of rains and the climate is semiarid, dry, sub-humid, and the soil is deep black, coastal alluvium. About 43 % of the area is under forests and a similar proportion is cultivated. Irrigation is spread over about 24 % of the cultivated area.

4.2.2 *Southern Gujarat*

This subzone, covering the districts of Surat and Bharuch, has seen rapid industrial development in the recent decade or so. Over half of the land is cultivated and about a fifth of the cultivated area is irrigated. The area receives little less than 974 mm of annual rainfall. The climate is semiarid, dry, subhumid, and the soil is deep black, coastal alluvium.

4.2.3 Middle Gujarat

Although this area is well developed industrially, it is also the most agrarian in Gujarat. Nearly two-thirds of the area is under cultivation and nearly a third of this is irrigated. Rains reduce progressively as one moves into North Gujarat. In the middle areas, which include Vadodara, Panch Mahals, and Kheda districts, the precipitation is of the order of 904 mm annually. The climate is semiarid and the soil is medium black.

4.2.4 North Gujarat

This subzone covers Banaskantha, Mehsana, Sabarkantha, Gandhinagar, and Ahmedabad districts. Land productivity is very low. Rainfall is only around 735 mm per annum. The climate is arid to semiarid and the soil is gray brown coastal alluvium. About 63 % of the area is cultivated and a little over a third of this is irrigated. The chief source of irrigation is ground water. However, in some areas, there is overdrawal of ground water.

4.2.5 Northwest Arid

This is the vast expanse of the Kachchh district. Rainfall is only about 340 mm per annum, the climate is arid to semiarid and the soil is gray brown, deltaic alluvium. Less than 13 % of the area is cultivated. Nearly one-third of the geographical area is wastelands.

4.2.6 North Saurashtra

This subzone includes the districts of Amreli, Bhavnagar, Jamnagar, Rajkot, and Surendranagar. The region receives 537 mm of rainfall and the climate is dry, subhumid. The soil is medium black calcareous. About 63 % of the area is cultivated, of which 24 % is irrigated. Agricultural productivity is relatively high in Saurashtra essentially because of the cultivation of groundnut in this region.

4.2.7 South Saurashtra

The South Saurashtra subzone includes only the district of Junagadh at the southwestern end of the state. This area receives little better rain than the nonsouth

Table 4.1 Selected agroclimatic features of subzones of Gujarat plains and hills region

No.	Subzone	Rainfall	Climate	Soil	Crops
1	Southern hills	1793	Semiarid to dry subhumid	Deep black, coastal alluvium	Rice, ragi, sugarcane, jowar
2	Southern Gujarat	974	Semiarid to dry subhumid	Deep black, coastal alluvium	Jowar, arhar, cotton, wheat
3	Middle Gujarat	904	Semiarid	Medium black	Rice, maize, bajra, cotton
4	South Saurashtra	844	Dry subhumid	Coastal alluvium, medium black	Groundnut, wheat, bajra, cotton
5	North Gujarat	735	Arid to semiarid	Gray brown, coastal alluvium	Bajra, cotton, jowar, wheat
6	North Saurashtra	537	Dry subhumid	Medium black	Bajra, jowar, groundnut, cotton
7	Northwest Arid	340	Arid to semiarid	Gray brown, deltaic alluvium	Bajra, groundnut, jowar, cotton

Gujarat parts of the state. The annual precipitation is about 844 mm, the climate is dry, subhumid, and the soil is coastal alluvium, medium black. About 56 % of the region is cultivated.

The selected agroclimatic features of subzones of Gujarat Plains and hills region are shown in Table 4.1.

4.3 Study Area

Gujarat is located on the west coast of India surrounded by the Arabian Sea in the west, Rajasthan in the north and northeast, Madhya Pradesh in the east, and Maharashtra in the south and southeast. It is situated between 20° 1′ and 24° 7′ north latitudes and 68° 4′ and 74° 4′ east longitudes.

Gujarat is a vibrant state in agricultural sector in terms of gross production of agricultural produce, productivity per hectare, adoption of new technology and innovations, crop diversification, introduction of new crops, postharvest technology and management. Gujarat has a diversified cropping pattern which includes the food grains and pulses, cash crops, and oil seeds. Major food grain crops are wheat, paddy, bajara, maize, etc., and pigeon pea, gram, greengram are the major pulses grown in the state. Cotton, castor, groundnut, mustard are the important oilseed crops of the state and the state has notable achievement in production and productivity scenario in cotton, castor, and groundnut. Cotton is an important crop of the state, which covers 26.33 lakh hectare area under cultivation and produced 98.25 lakh bales during 2010–11, which is 1/3rd production of the country. This state has recognition for highest productivity in the world for castor which is 1984 kg/ha. This state produced 84 % of the total castor production of the country, with an area of 4.91 lakh hectare and 9.71 lakh MT production. This state has a 30 % share in the country for production of groundnut, with 33.76 lakh MT production through area coverage of 18.05 lakh hectare and has achieved 100 lakh MT food grain production first time in a year, too.

The entire state of Gujarat is divided into the various agroclimatic zones. Vallabh Vidyanagar is located in the Anand district and lies in Middle Gujarat agroclimatic zone-III of Gujarat state. Vallabh Vidyanagar is located at 22° 32′ N latitude, 72° 54′ E longitude at an altitude of 34 m above mean sea level. It is bounded on the north by the Kheda district, on the south by the Gulf of Khambhat, on the West by Ahmedabad district, and on the East by Vadodara district. The climate of Vallabh Vidyanagar is semiarid with fairly dry and hot summer. Winter is fairly cold and sets in, in the month of November and continues till the middle of February. Summer is hot and dry which commences from mid of February and ends by the month of June. May is the hottest month with mean maximum temperature around 40.08 °C. The average rainfall is 853 mm. The soil of the region is popularly known as Goradu soil. It is alluvial in origin. The texture of the soil is sandy loam and black. The soil is deep enough to respond well to anuring and variety of crops of the tropical and subtropical regions. The soil is low in organic carbon and nitrogen, medium in available phosphorus and available sulfur. In this area paddy, tur, cotton, groundnut, and til are grown in kharif season. In Rabi season wheat, gram, and jowar are grown. Especially, in summer season bajara and groundnut are grown. Tobacco is grown from August and harvested in March. In last few years, there is an increase in amount of rainfall which facilitated in agriculture production and various irrigation schemes.

4.4   Data Collection

The data required for study are collected from state water data centre and Krishi Bhavan, Gandhinagar. Long term climatological yearly data are collected for Vallabh Vidyanagar, Anand district of Gujarat. The basic annual climatological data obtained comprises of maximum and minimum temperature (°C), relative humidity (%), wind speed (Kmph), and sunshine hours (hours). The yield data of various crops grown in Vallabh Vidyanagar are collected from the Krishi Bhavan, Gandhinagar from 1981 to 2006.

Chapter 5
Methodology

Abstract This chapter gives the overview on the methodology to predict the yield of cotton using multiple linear regression and principal component regression, which is followed by models' performance indices to estimate the best model. The climatological parameters considered to predict the yield of cotton were maximum temperature, minimum temperature, wind velocity, relative humidity, and sunshine hours. The methodology of MLR and PCR models to predict the yield of cotton, considering the above-given climatological parameters, as input and yield of cotton as output, is discussed. The model's performances were evaluated for training and validation (70–30 %) using performance indices such as root mean squared error (RMSE), coefficient of correlation (r), coefficient of determination (R^2), and discrepancy ratio (D.R.).

Keywords MLR model methodology · PCR model methodology · Performance indices

5.1 Multiple Linear Regression Model

Multiple linear regression is an extension of simple linear regression in which more than one independent variable is used to predict single dependent variable "Y." The predicted value of "Y" is a linear transformation of the variables such that the sum of squared deviations of the observed and predicted "Y" is a minimum. The computations are more complex, however, because the interrelationships among all variables must be taken into account in the weights assigned to the variables. Here maximum temperature, minimum temperature, wind velocity, relative humidity, and sunshine hours are selected as independent variables x_1, x_2, x_3, x_4, x_5 and yield of cotton as dependent variable y. The equation will be of the form as given below.

$$\mathbf{y} = \mathbf{d_0} + \mathbf{d_1 x_1} + \mathbf{d_2 x_2} + \mathbf{d_3 x_3} + \mathbf{d_4 x_4} + \mathbf{d_5 x_5} \tag{5.1}$$

T.M.V. Suryanarayana and P.B. Mistry, *Principal Component Regression for Crop Yield Estimation*, SpringerBriefs in Applied Sciences and Technology, DOI 10.1007/978-981-10-0663-0_5

The coefficients in the above-given equation are obtained by multiple linear regression which are thereafter put in Eq. (5.1) to estimate the value of y for known values of independent variables.

5.2   Principal Component Regression Model

The variables under study are highly correlated. It may be useful to transform the original set of variables to a new set of uncorrelated variables called principal components. These new variables are linear combinations of original variables and are derived in decreasing order of importance, so that the first principal component accounts for as much as possible of the variation in the original data.

The first few principal components account for most of the variability in the original data. These few principal components can then replace the initial p variables in subsequent analysis and thus, reducing the effective dimensionality of the problem. Further, since Principal Component Analysis transforms original set of variables to new set of uncorrelated variables, it is worth stressing that if original variables are uncorrelated, then, there is no point in carrying out principal component analysis.

In order to find principal components, the covariance matrix of the selected data is generated. Also, eigenvalues and eigenvectors of covariance matrix are found and principal components of the data considered are derived.

It can be seen that the total variation explained by principal components is same as that explained by original variables. It could also be proved mathematically as well as empirically that the principal components are uncorrelated. The proportion of total variation accounted for, by the first principal component, z_1 is

$$\frac{\lambda1}{\lambda1 + \lambda2 + \lambda3 + \lambda4 + \lambda5 + \lambda6}$$

The proportion of total variation accounted for, by the second principal component, z_2 is

$$\frac{\lambda1 + \lambda2}{\lambda1 + \lambda2 + \lambda3 + \lambda4 + \lambda5 + \lambda6}$$

Hence, the first or first two principal components, i.e., z_1 and z_2, could replace variables which are maximum temperature, minimum temperature, relative humidity, wind speed, and sunshine hours by sacrificing negligible information about the total variation in the system. Thus, the whole data will be converted to a new data set with two principal components. The larger eigenvalue is associated with the first principal component. The next larger eigenvalue is associated with the second principal component.

The derived principal components z_1 and z_2, are then considered as regressors and the dependent variable as regressand. The principal component regression is carried out by using the above said regressors and regressand in curve expert, and a series of regression models are developed. The model with maximum correlation coefficient and minimum standard error and hence having placed at Rank 1, is chosen as the best model.

5.3 Performance Indices

The results of the model developed in this study were evaluated by means of following performance indices:

5.3.1 Root Mean Squared Error (RMSE)

The root mean squared error (RMSE) (also called the root mean square deviation, RMSD) is a frequently used measure of the difference between values predicted by a model and the values actually observed from the environment that is being modelled. These individual differences are also called residuals, and the RMSE serves to aggregate them into a single measure of predictive power.

The RMSE of a model prediction with respect to the estimated variable X_{model} is defined as the square root of the mean squared error. The equation will be of the form as given below in Eq. (5.2).

$$\text{RMSE} = \sqrt{\frac{\sum_{i=1}^{n}\left(X_{obs} - X_{model}\right)^2}{n}} \tag{5.2}$$

where X_{obs} is observed values and X_{model} is modelled values at time/place i. Also n is the number of observations.

However, the RMSE values can be used to distinguish model performance in a calibration period with that of a validation period as well as to compare the individual model performance to that of other predictive models.

5.3.2 Correlation Coefficient (r)

Correlation, often measured as a correlation coefficient, indicates the strength and direction of a linear relationship between two variables (for example, model output and observed values). A number of different coefficients are used for different situations. The best known is the Pearson product-moment correlation coefficient (also called Pearson correlation coefficient or the sample correlation coefficient), which is obtained by dividing the covariance of the two variables by the product of their standard deviations. If one has a series of n observations and n model values, then the Pearson product-moment correlation coefficient can be used to estimate the correlation between model and observations. The equation will be of the form as given below in Eq. (5.3).

$$r = \frac{\sum_{i=1}^{n}(x_i - \bar{x})(y_i - \bar{y})}{\sqrt{\sum_{i=1}^{n}(x_i - \bar{x})^2 \sum_{i=1}^{n}(y_i - \bar{y})^2}} \qquad (5.3)$$

where x_i is observed values and y_i is modelled values at time/place i, $\bar{x}$ and $\bar{y}$ is the average value of observed and predicted values.

The correlation is +1 in the case of a perfect increasing linear relationship, and −1 in case of a decreasing linear relationship, and the values in-between indicates the degree of linear relationship between model and observations. A correlation coefficient of zero means that there is no linear relationship between the variables.

5.3.3   Coefficient of Determination (R^2)

A measure used in statistical model analysis to assess how well a model explains and predicts future outcomes. It is indicative of the level of explained variability in the model. The coefficient, also commonly known as R-square, is used as a guideline to measure the accuracy of the model. The equation will be of the form as given below in Eq. (5.4).

$$R^2 = \left(\frac{\sum_{i=1}^{n}(y_i - \bar{y})(\hat{y}_i - \tilde{y})}{\sqrt{\sum_{i=1}^{n}(y_i - \bar{y})^2 \sum_{i=1}^{n}(\hat{y}_i - \tilde{y})^2}} \right)^2 \qquad (5.4)$$

5.3.4   Discrepancy Ratio (D.R.)

It is the ratio of observed values (y_i) and predicted values ($\bar{y}_i$) of the dependent variables. The equation will be of the form as given below in Eq. (5.5).

$$D.R. = \frac{\sum_{i=1}^{n} y_i}{\sum_{i=1}^{n} \bar{y}_i} \qquad (5.5)$$

5.4   Analysis of MLR and PCR Models

The whole data set has been divided into 70–30 %. That means 70 % data has been considered for training and the remaining 30 % data has been selected for validation of the models.

- The root mean squared error value obtained during training and validation should be almost equal and it should be numerically as minimum as possible.
- The coefficient of correlation should be as maximum as possible in the range of 0 to 1, and preferably above 0.80 and as possible as nearer to unity.
- The coefficient of determination should be nearer to unity.
- The discrepancy ratio should also be nearer to unity.

The models were evaluated using the above performance indices and the best model will be concluded and recommended for estimation of the dependent variable.

Chapter 6
Results and Analysis

Abstract This chapter illustrates the results and analysis of the study. The MLR models are developed and the results are given. The performance indices for the MLR model are analyzed. This is followed by the development of PCR model and the same are analyzed using the above said indices. The outcomes are statistically analyzed and their accuracy are assessed and discussed during training and validation.

Keyword Trained MLR model · Validated MLR model · Trained PCR model · Validated PCR model

6.1 MLR Model During Training and Validation

Multiple linear regression is done using maximum temperature, minimum temperature, relative humidity, sunshine hours, and wind speed as independent variables and crop yield as dependent variable.

The generalized form of multiple linear regression is as follows:

$$\mathbf{y} = \mathbf{d}_0 + \mathbf{d}_1\mathbf{x}_1 + \mathbf{d}_2\mathbf{x}_2 + \mathbf{d}_3\mathbf{x}_3 + \mathbf{d}_4\mathbf{x}_4 + \mathbf{d}_5\mathbf{x}_5 \tag{6.1}$$

The coefficients of MLR model developed during training, i.e., 70 % of dataset considered, are shown in Table 6.1.

6.1.1 Multiple Linear Regression During Training

The results of MLR model for observed and predicted yield of cotton during training are shown in Table 6.2.

The results of the performance indices, viz., root mean squared error (RMSE), coefficient of correlation (r), coefficient of determination (R^2), and discrepancy ratio (DR) for the MLR model developed during training are given below:

© The Author(s) 2016
T.M.V. Suryanarayana and P.B. Mistry, *Principal Component Regression for Crop Yield Estimation*, SpringerBriefs in Applied Sciences and Technology, DOI 10.1007/978-981-10-0663-0_6

Table 6.1 Coefficients of MLR model developed

Coefficients	Value
d_0	114.5882
d_1	5.52183
d_2	11.88686
d_3	2.178833
d_4	−74.7393
d_5	−9.44541

Table 6.2 Observed and predicted yield of cotton using MLR during training

Year	Observed yield (Kg/ha)	Predicted yield (Kg/ha)
1981	170	233.02
1982	175	210.80
1983	179	205.89
1984	200	263.53
1985	263	267.33
1986	275	255.63
1987	290	231.57
1988	305	331.09
1989	319	284.30
1990	326	265.25
1991	218	305.24
1992	274	246.68
1993	277	232.59
1994	340	298.00
1995	300	304.99
1996	293	334.21
1997	323	331.15
1998	410	332.70

$$RMSE = 46.6989 \, kg/ha, \; r = 0.6664, \; R^2 = 0.4440 \; and \; D.R. = 0.9999$$

The comparison of observed yield of cotton and predicted yield of cotton using MLR model during training is shown in Figs. 6.1 and 6.2.

6.1.2   *Multiple Linear Regression During Validation*

The results of MLR model for observed and predicted yield of cotton during validation are shown in Table 6.3.

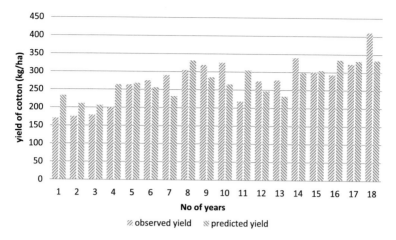

Fig. 6.1 Observed and predicted yield of cotton using MLR during training

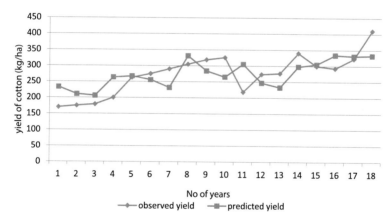

Fig. 6.2 Scatter plot of observed and predicted yield of cotton using MLR during training

Table 6.3 Observed and predicted yield of cotton using MLR during validation

Years	Observed yield (Kg/ha)	Predicted yield (Kg/ha)
1999	264	294.41
2000	219	340.02
2001	205	318.94
2002	210	341.37
2003	220	258.97
2004	265	353.46
2005	339	388.71
2006	232	405.92

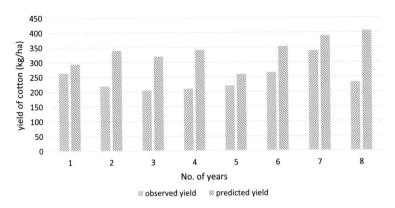

Fig. 6.3 Observed and predicted yield of cotton using MLR during validation

The results of the performance indices, viz., root mean squared error (RMSE), coefficient of correlation (r), coefficient of determination (R^2), and discrepancy ratio (DR) for the developed MLR model during validation are given below:

$$RMSE = 104.80 \, kg/ha, r = 0.3985, R^2 = 0.1588 \text{ and D.R.} = 0.7232$$

The comparison of observed yield of cotton and predicted yield of cotton using MLR model during validation is shown in Figs. 6.3 and 6.4.

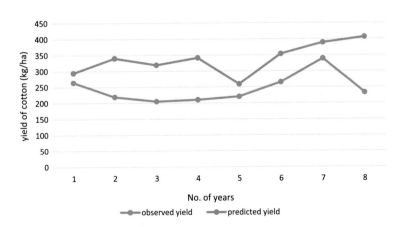

Fig. 6.4 Scatter plot of observed and predicted yield of cotton using MLR during validation

6.2   PCR Model During Training and Validation

The variables under study are highly correlated. So it may be useful to transform the dataset of variables to a new set of uncorrelated variables called principal components. These new variables are linear combinations of original variables and are derived in decreasing order of importance so that the first principal component accounts for as much as possible of the variation in the original data.

Now, the variance–covariance matrix is as follows:

Covariance matrix:

$$
\begin{bmatrix}
0.7991 & 0.3548 & 0.2052 & 0.1563 & -0.083 & 4.8282 \\
0.3548 & 0.5392 & 1.4073 & 0.1798 & -0.292 & 9.2796 \\
0.2052 & 1.4073 & 32.204 & 1.8038 & -1.656 & 50.599 \\
0.1563 & 0.1798 & 1.8038 & 0.6105 & -0.18 & -9.34 \\
-0.083 & -0.292 & -1.656 & -0.18 & 0.4376 & -8.773 \\
4.8282 & 9.2796 & 50.599 & -9.34 & -8.773 & 3442.3
\end{bmatrix}
$$

The eigenvalues (λ) and eigenvectors (a_i) of the above matrix are obtained for given dataset and are given as follows in decreasing order along with the corresponding eigenvectors.

Eigenvalues (λ):

$$
\begin{bmatrix}
3443.2 \\
31.7 \\
1.1 \\
0.5 \\
0.4 \\
0.1
\end{bmatrix}
$$

Eigenvectors (a_i):

$$
\begin{bmatrix}
0.0014 & 0.0027 & 0.0148 & -0.0027 & -0.003 & 0.9999 \\
0.0052 & 0.414 & 0.9958 & 0.0628 & -0.049 & -0.015 \\
0.7495 & 0.5218 & -0.058 & 0.3218 & -0.243 & -0.001 \\
0.5632 & -0.188 & 0.0687 & -0.5673 & 0.567 & -0.001 \\
0.1475 & -0.465 & -0.006 & 0.7484 & 0.45 & 0.0043 \\
0.3151 & -0.689 & 0.0016 & -0.1027 & -0.644 & -0.0005
\end{bmatrix}
$$

Therefore, the above eigenvalues and eigenvectors may now be represented as below.

$$\lambda_1 = 3443.2, \ \lambda_2 = 31.7, \ \lambda_3 = 1.1, \ \lambda_4 = 0.5, \ \lambda_5 = 0.4, \ \lambda_6 = 0.1$$

$$a_1 = (0.0014, 0.0027, 0.0148, -0.0027, -0.0030, 0.9999)$$

$$a_2 = (0.0052, 0.4140, 0.9958, 0.0628, -0.0490, -0.0150)$$

$$a_3 = (0.7495, 0.5218, -0.0580, 0.3218, -0.2430, -0.0010)$$

$$a_4 = (0.5632, -0.1880, 0.0687, -0.5673, 0.5670, -0.0010)$$

$$a_5 = (0.1475, -0.4650, -0.0060, 0.7484, 0.4500, 0.0043)$$

$$a_6 = (0.3151, -0.6890, 0.0016, -0.1027, -0.6440, -0.0005)$$

The principal components for data will be

$$z_1 = 0.0014x_1 + 0.0027x_2 + 0.0148x_3 - 0.0027x_4 - 0.003x_5 + 0.9999x_6$$

$$z_2 = 0.0052x_1 + 0.4140x_2 + 0.9958x_3 + 0.0628x_4 - 0.0490x_5 - 0.0150x_6$$

$$z_3 = 0.7495x_1 + 0.5218x_2 - 0.0580x_3 + 0.3218x_4 - 0.2430x_5 - 0.0010x_6$$

$$z_4 = 0.5632x_1 - 0.1880x_2 + 0.0687x_3 - 0.5673x_4 + 0.5670x_5 - 0.0010x_6$$

$$z_5 = 0.1475x_1 - 0.4650x_2 - 0.0060x_3 + 0.7484x_4 + 0.4500x_5 + 0.0043x_6$$

$$z_6 = 0.3151x_1 - 0.6890x_2 + 0.0016x_3 - 0.1027x_4 - 0.6440x_5 - 0.0005x_6$$

The variance of principal components will be eigenvalues, i.e.,

$$\text{var}(z_1) = 3443.2, \ \text{var}(z_2) = 31.7, \ \text{var}(z_3) = 1.1, \ \text{var}(z_4) = 0.5,$$
$$\text{var}(z_5) = 0.4, \ \text{var}(z_6) = 0.1$$

The total variation explained by original variables is

$$\text{var}(x_1) + \text{var}(x_2) + \text{var}(x_3) + \text{var}(x_4) + \text{var}(x_5) + \text{var}(x_6) = 3476.90$$

The total variation explained by principal components is

$$\lambda_1 + \lambda_2 + \lambda_3 + \lambda_4 + \lambda_5 + \lambda_6 = 3477$$

As such, it can be seen that the total variation explained by principal components is same as that explained by original variables. It could also be proved mathematically as well as empirically that the principal components are uncorrelated. The proportion of total variation accounted for by the first principal component is

$$\frac{\lambda_1}{\lambda_1 + \lambda_2 + \lambda_3 + \lambda_4 + \lambda_5 + \lambda_6} = 0.9902$$

Continuing, the first two components account for a proportion

$$\frac{\lambda_1 + \lambda_2}{\lambda_1 + \lambda_2 + \lambda_3 + \lambda_4 + \lambda_5 + \lambda_6} = 1.0020$$

of the total variance.

Hence, the first or first two principal components, i.e., z_1 and z_2, could replace variables which are maximum temperature, minimum temperature, relative humidity, wind speed, and sunshine hours by sacrificing negligible information about the total variation in the system. The scores of principal components can be obtained by substituting the values of x_i in equations of z_i. For the dataset, the first two principal components for the year 1981, are given in Eqs. (6.2) and (6.3).

$$z_1 = 0.0014(38.2) + 0.0027(22.55) + 0.0148(51.15) - 0.0027(4.88) \\ - 0.003(11.34) + 0.9999(170) = 170.81 \tag{6.2}$$

$$z_2 = 0.0052(36.05) + 0.414(21.67) + 0.9958(52.5) + 0.0628(4.88) \\ - 0.049(11.64) - 0.015(175) = 57.70 \tag{6.3}$$

Thus, the whole data has been converted to a new dataset with two principal components. The larger eigenvalue is associated with the first principal component. The next larger eigenvalue is associated with the second principal component.

Now, principal component regression (PCR) is a type of regression analysis, which considers principal components (PCs) as independent variables instead of adopting original variables. The PCs are the linear combination of the original variables, which can be obtained by principal component analysis (PCA).

The derived principal components z_1 and z_2, are then considered as regressors and the dependent variable as regressand. The principal component regression is carried out using the above said regressors and regressand in curve expert, and a series of regression models are developed. The model with maximum correlation coefficient and minimum standard error and hence having placed at Rank 1, is chosen as the best model. The best model developed for the area considered for the study is Power Model B, which is of the form as given in Eq. 6.4.

$$\mathbf{y = a + z_1^b + z_2^c} \tag{6.4}$$

where a, b, c are coefficients and their values

$$a = -1.164, \ b = 0.9999 \text{ and } c = -0.2493$$

Table 6.4 Observed and predicted yield of cotton using PCR during training

Year	Observed yield (Kg/ha)	Predicted yield (Kg/ha)
1981	170	169.92
1982	175	174.93
1983	179	178.84
1984	200	199.80
1985	263	262.81
1986	275	274.92
1987	290	289.87
1988	305	304.91
1989	319	318.86
1990	326	325.94
1991	218	217.91
1992	274	273.81
1993	277	276.80
1994	340	339.72
1995	300	299.80
1996	293	292.92
1997	323	322.79
1998	410	409.64

6.2.1   Principal Component Regression During Training

The results of observed and predicted yield of cotton using PCR Model during training are shown in Table 6.4.

The results of the performance indices, viz., root mean squared error (RMSE), coefficient of correlation (r), coefficient of determination (R^2), and discrepancy ratio (DR) during training are shown below:

$$RMSE = 0.1745 \, kg/ha, \; r = 0.9999, \; R^2 = 0.9998 \; and \; D.R. = 1.0000$$

The comparison of observed yield of cotton and predicted yield of cotton using PCR model during training is illustrated in Figs. 6.5 and 6.6.

6.2.2   PCR During Validation

The results of observed and predicted yield of cotton using PCR Model during validation are shown in Table 6.5.

The results of the performance indices, viz., root mean squared error (RMSE), coefficient of correlation (r), coefficient of determination (R^2), and discrepancy ratio (DR) for the developed PCR model during validation are shown below:

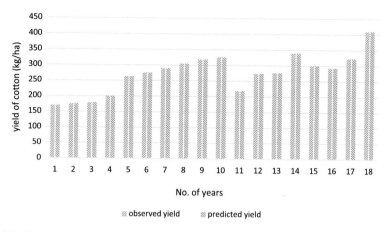

Fig. 6.5 Observed and predicted yield of cotton using PCR during training

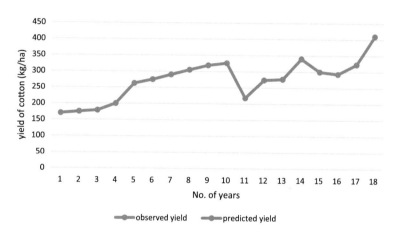

Fig. 6.6 Scatter plot of observed and predicted yield of cotton using PCR during training

Year	Observed yield (Kg/ha)	Predicted yield (Kg/ha)
1999	264	263.68
2000	219	218.91
2001	205	204.75
2002	210	209.81
2003	220	219.79
2004	265	264.89
2005	339	338.73
2006	232	231.81

Table 6.5 Observed and predicted yield of cotton using PCR during validation

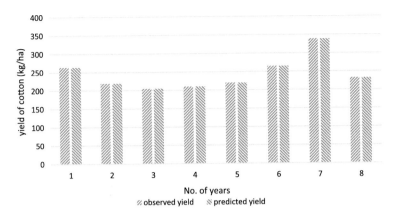

Fig. 6.7 Observed and predicted yield of cotton using PCR during validation

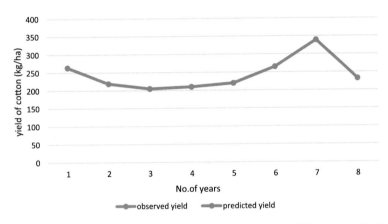

Fig. 6.8 Scatter plot of observed and predicted yield of cotton using PCR during validation

$$\text{RMSE} = 0.2140\,\text{kg/ha},\ r = 0.9999,\ R^2 = 0.9998\ \text{and D.R.} = 0.9999$$

The comparison of observed yield of cotton and predicted yield of cotton using PCR model during validation is shown in Figs. 6.7 and 6.8.

6.3   Comparison of MLR and PCR Models Using Performance Indices

Comparisons of performance indices using the developed MLR and PCR models during training and validation are given in Tables 6.6 and 6.7, respectively.

Table **6.6** Comparison of results of MLR and PCR model during training	Performance indices (Training-70 %)			
	RMSE	r	R^2	D.R.
MLR	46.6920	0.6664	0.4435	0.9999
PCR	0.1745	0.9999	0.9998	1.000

Table **6.7** Comparison of results of MLR and PCR models during validation	Performance indices (Validation-30 %)			
	RMSE	r	R^2	D.R.
MLR	104.8000	0.3985	0.1588	0.7232
PCR	0.2140	0.9999	0.9998	0.9999

6.4 Analysis of MLR and PCR Models Developed

The model developed using multiple linear regression for estimation of yield of cotton gives RMSE values as 46.69 kg/ha and 104.80 kg/ha, r value as 0.66 and 0.39, R^2 value as 0.44 and 0.15, and DR value as 0.99 and 0.72 during training and validation, respectively.

The model developed using principal component regression for estimation of yield of cotton gives RMSE values as 0.17 kg/ha and 0.21 kg/ha, r value as 0.99 and 0.99, R^2 value as 0.99 and 0.99, and DR value as 1.00 and 0.99 during training and validation, respectively.

The root mean squared error (RMSE) values by PCR model, when compared by those obtained by MLR model, decreased by almost 100 %, both during training and validation.

The coefficient of correlation (r) values by PCR model, when compared by those obtained by MLR model, have increased from 0.66 to 0.99 and 0.39 to 0.99 during training and validation, respectively.

The coefficient of determination (R^2) values by PCR model, when compared by those obtained by MLR model, have increased from 0.44 to 0.99 and 0.15 to 0.99 during training and validation, respectively.

The discrepancy ratio (DR) values by PCR model, when compared by those obtained by MLR model, are also nearer to unity during training and validation.

Chapter 7
Conclusions

Abstract This chapter summarizes the results obtained during the study, and the comparisons of the results by MLR and PCR are focused and the conclusions of the study are drawn out.

Keywords Estimated yield · Performance indices' values

7.1 Conclusions Based on the Study

The estimation of yield of cotton using climatic parameters such as maximum temperature, minimum temperature, relative humidity, wind speed, and sunshine hours is carried out using multiple linear regression and principal component regression.

The model developed using multiple linear regression for estimation of yield of cotton gives RMSE values as 46.6989 kg/ha and 104.8000 kg/ha, r value as 0.6664 and 0.3985, R^2 value as 0.4435 and 0.1588 and D.R. value as 0.9999 and 0.7232 during training and validation, respectively.

The model developed using principal component regression for estimation of yield of cotton gives RMSE values as 0.1745 kg/ha and 0.2140 kg/ha, r value as 0.9999 and 0.9999, R^2 value as 0.9998 and 0.9998 and D.R. value as 1.0000 and 0.9999 during training and validation, respectively.

The root mean squared error (RMSE) values by PCR model, when compared by those obtained by MLR model, decreased by almost 100 %, both during training and validation.

The coefficient of correlation (r) values by PCR model, when compared by those obtained by MLR model, have increased from 0.66 to 0.99 and 0.39 to 0.99 during training and validation, respectively.

The coefficient of determination (R^2) values by PCR model, when compared by those obtained by MLR model, have increased from 0.44 to 0.99 and 0.15 to 0.99 during training and validation respectively.

© The Author(s) 2016
T.M.V. Suryanarayana and P.B. Mistry, *Principal Component Regression for Crop Yield Estimation*, SpringerBriefs in Applied Sciences and Technology, DOI 10.1007/978-981-10-0663-0_7

The discrepancy ratio (D.R.) values by PCR model, when compared by those obtained by MLR model, are also nearer to unity during training and validation.

Hence, the principal component regression yields the better model, for the estimation of yield of cotton, for the area considered for the study.

References

Aksornsingchai P, Srinilta C (2011) Statistical downscaling for rainfall and temperature prediction in Thailand. The International Multiconference of Engineers and Computer Scientists (IMECS 2011), Vol 1, ISSN:2078-0966

Aydinalp C, Cresser C (2008) The effects of global climate change on agriculture. American-Eurasian J Agric Environ Sci 3(5):672–676. ISSN:1818-6769

Bates BC, Kundzew ZW, Palutikof JP(eds) (2008) IPCC climate change and water. Technical paper of the intergovernmental panel on climate change, IPCC Secretariat, Geneva, pp 210

Chen X, Chen C, Jin L (2011) Principal component analyses in anthropological genetics. Adv in Anthropol 1(2):9–14

Devak M, Dhanya CT (2014) Downscaling of precipitation in Mahanadi basin, India. Int J Civil Eng Res 5(2):111–120. ISSN:2278-3652

Dutta PS, Tahbilder H (2014) Prediction of rainfall using data mining technique over Assam. Indian Journal of Computer Science and Engineering (IJCSE) 5(2):85–90. ISSN:0976-5166

Fekedulegn BD, Colbert JJ, Hicks R, Schuckers ME (2002) Coping with multicollinearity: an example on application of principal components regression in dendroecology. research paper NE-721, USDA forest service

Haque MM, Rahman A, Hagare D, Kibria G (2013) Principal component regression analysis in water demand forecasting: an application to the Blue mountains, NSW, Australia. J Hydrol Environ Res 1(1):49–59

Joilliffe IT (2002) Principal component analysis, 2nd edn. Springer Series in Statistics, Springer, Heidelberg

Kellow J (2006) Using principal components analysis in program evaluation: some practical considerations. J Multi Eval 5:89–107. ISSN:1556-8180

Kumar A, Sharma P (2013) Impact of climate variation on agricultural productivity and food security in rural India. Economics discussion paper, 2013–2043, August 15

Kumar S, Chauhan A (2014) A survey on image feature selection techniques. International Journal of Computer Science and Information Technologies 5(5):6449–6452

Lee J, Six J (2010) Effect of climate change on field crop production and greenhouse gas emissions in California's central valley. 19th World Congress of Soil Science, soil solutions for a changing world, Brisbane, Australia, 1–6 August 2010, pp 52–55

Li-Jun F (2009) Statistically downscaled temperature scenarios over China. Atmos Oceanic Sci Lett 2(4):208–213

Ma S (2007) Principal component analysis in linear regression survival model with microarray data. J Data Sci, pp 183–198

Mendes M (2011) Multivariate multiple regression analysis based on principal component scores to study relationships between some pre- and post-slaughter traits of broilers. J Agri Sci 17:77–83

© The Author(s) 2016
T.M.V. Suryanarayana and P.B. Mistry, *Principal Component Regression for Crop Yield Estimation*, SpringerBriefs in Applied Sciences and Technology, DOI 10.1007/978-981-10-0663-0

Mustapha A, Abdu A (2012) Application of principal component analysis & multiple regression models in surface water quality assessment. J Environ Earth Sci 2(2):16–24. ISSN:2224-3216, ISSN:2225-0948

Ojha CSP, Goyal MK, Adeloye AJ (2010) Downscaling of precipitation for lake catchment in arid region in India using linear multiple regression and neural networks. J Open Hydrol 4:122–136

Park B, Chen YR, Hruschka WR, Shackelford SD, Koohmaraie M (2001) Principal component regression of near-infrared reflectance spectra for beef tenderness prediction. Trans Am Soc Agric Eng 44(3):609–615

Peres-Neto PR, Jackson DA, Somers KM (2005) How many principal components? stopping rules for determining the number of non-trivial axes revisited. Comput Stat Data Anal 49:974–997

Ramasamy R, Swamy A (2012) Global warming, climate change and tourism: a review of literature. Special issue: sustainability, tourism & environment in the shift of a millennium: a peripheral view, Year 6, No. 03. pp 72–98

Ramesh D, Vardhan BV (2015) Analysis of crop yield prediction using data mining techniques. IJRET: International Journal of Research in Engineering and Technology 4(1):470–473, Jan 2015. eISSN:2319-1163, pISSN:2321-7308

Schoof JT, Pryor SC (2001) Downscaling temperature and precipitation: a comparison of regression-based methods and artificial neural networks. Int J Climatol 21:773–790

Tatli H, Mentes S (2004) A statistical downscaling method for monthly total precipitation over Turkey. Int J Climatol 24:161–180

Tisseuil C, Vrac M, Lek S, Wade AJ (2010) Statistical downscaling of river flows. J Hydrol 385:279–291

Trigo RM, Palutikof JP (1999) Simulation of daily temperatures for climate change scenarios over Portugal: a neural network model approach. Climate research, vol 13, 7 Sept, pp 45–59

Ul-Saufie AZ, Yahya AS, Ramali NA (2011) Improving multiple linear regression model using principal component analysis for predicting PM10 concentration in Seberang Prai, Pulau Pinang. Int J Environ Sci 2(2):403–410

Webster TJ (2001) A principal component analysis of the U.S. news & world report tier rankings of colleges and universities. Economics of Education Review, pp 235–244

Yadav RR, Sisodia BVS, Kumar S (2014) Application of principal component analysis in developing statistical models to forecast crop yield using weather variables. 631.153.028: 551.506.1. MAUSAM 65:357–360

Yan G, Ping LJ, Yun L (2011) Statistically downscaled summer rainfall over the middle-lower reaches of the Yangtze river. Atmos Oceanic Sci Lett 4:191–198

Online Documents

A.2: FAQ on Science of Climate Change http://ec.gc.ca/scitech/default.asp?lang=En&n= 2A953C90-1&offset=1&toc=show

Agrawal R, Rao AR, Data reduction techniques. http://www.iasri.res.in/ebook/EB_SMAR/e-book_pdf%20files/Manual%20II/9-data_reduction.pdf. Accessed 25 Dec 2015

Agro climatic zones in India. http://vikaspedia.in/agriculture/crop-production/weather-information/ agro-climatic-zones-in-india. Accessed 1 Jan 2016

Bareja BG (2011) Climatic factors can promote or inhibit plant growth and development. http:// www.cropsreview.com/. Accessed 18 Oct 2015

Barrow E (2001) The availability, characteristics and use of climate change scenarios PARC workshop. http://www.parc.ca/pdf/conference_proceedings/jan_01_barrow4.pdf. Accessed 2 Dec 2015

Ben, Bareja (2011) Climatic factors can promote or inhibit plant growth and development. http:// Www.Cropsreview.Com/Climatic-Factors.Html. Accessed 18 Oct 2015

Climate Change 2007: working group I: the physical science basis, FAQ 1.2. https://www.ipcc.ch/ publications_and_data/ar4/wg1/en/faq-1-2.html

Climate change adaptation: the pivotal role of water, UN water policy brief. http://www.unwater.
org/downloads/unw_ccpol_web.pdf

Gautam (2010) Climate change scenario generation using statistical downscaling. http://
Shodhbhagirathi.Iitr.Ac.In:8081/Jspui/Handle/123456789/3197. Accessed 15 Nov 2015

Gujarat State Farmer Guide (2011) Mechanization and Technology Division, Ministry of
Agriculture, Department of Agriculture and cooperation, Government of India http://farmech.
dac.gov.in/FarmerGuide/GJ/index1.html. Accessed 08 Jan 2016

Harun (2008) Regional climate scenarios using a statistical downscaling approach. http://core.ac.
uk/download/files/392/11783814.pdf. Accessed 1 Jun 2015

HGIC 1050, http://www.clemson.edu/extension/hgic/plants/pdf/hgic1050.pdf

Introduction to Applied Statistics, Lesson: Multiple Linear Regression. http://sites.stat.psu.edu/
~ajw13/stat200/mos/12_multregr/12_multregr_print.html

Multiple Regression analysis. http://www.unt.edu/rss/class/mike/5710/Multiple%20Regression.
pdf. Accessed 25 Nov 2015

Panjabsingh, Agro-Climatic Zonal Planning Including Agriculture Development In North-Eastern
India.   http://planningcommission.nic.in/aboutus/committee/wrkgrp11/wg11_agrclim.pdf.
Accessed 1 Jan 2016

Pathak, Long-term strategies and programmes for mechanization of agriculture in agro climatic
zone–xiii: Gujarat plains and hills region. http://farmech.dac.gov.in/06035-04-ACZ13-
15052006.pdf. Accessed 20 Aug 2015

Polynomial Regression model. http://home.iitk.ac.in/~shalab/regression/Chapter12-Regression-
PolynomialRegression.pdf. Accessed 25 Nov 2015

Solanki (2012) Gujarat agriculture. http://nfsm.gov.in/blog/gujarat_agriculture.aspx. Accessed 25
Dec 2015

TNAU Agricultural Portal, http://agritech.tnau.ac.in/agriculture/agri_agrometeorology_wind.html

Trzaska S, Schnarr E (2014) A review of downscaling methods for climate change projections.
http://www.ciesin.org/documents/Downscaling_CLEARED_000.pdf. Accessed 25 Nov 2015

Printed in the United States
By Bookmasters